Muir Kumph

2D Arrays of Ion Traps for Quantum Information Processing

Muir Kumph

2D Arrays of Ion Traps for Quantum Information Processing

Sculpting the Pseudopotential Landscape of Trapped Ions

Südwestdeutscher Verlag für Hochschulschriften

Impressum / Imprint

Bibliografische Information der Deutschen Nationalbibliothek: Die Deutsche Nationalbibliothek verzeichnet diese Publikation in der Deutschen Nationalbibliografie; detaillierte bibliografische Daten sind im Internet über http://dnb.d-nb.de abrufbar.

Alle in diesem Buch genannten Marken und Produktnamen unterliegen warenzeichen-, marken- oder patentrechtlichem Schutz bzw. sind Warenzeichen oder eingetragene Warenzeichen der jeweiligen Inhaber. Die Wiedergabe von Marken, Produktnamen, Gebrauchsnamen, Handelsnamen, Warenbezeichnungen u.s.w. in diesem Werk berechtigt auch ohne besondere Kennzeichnung nicht zu der Annahme, dass solche Namen im Sinne der Warenzeichen- und Markenschutzgesetzgebung als frei zu betrachten wären und daher von jedermann benutzt werden dürften.

Bibliographic information published by the Deutsche Nationalbibliothek: The Deutsche Nationalbibliothek lists this publication in the Deutsche Nationalbibliografie; detailed bibliographic data are available in the Internet at http://dnb.d-nb.de.

Any brand names and product names mentioned in this book are subject to trademark, brand or patent protection and are trademarks or registered trademarks of their respective holders. The use of brand names, product names, common names, trade names, product descriptions etc. even without a particular marking in this work is in no way to be construed to mean that such names may be regarded as unrestricted in respect of trademark and brand protection legislation and could thus be used by anyone.

Coverbild / Cover image: www.ingimage.com

Verlag / Publisher:
Südwestdeutscher Verlag für Hochschulschriften
ist ein Imprint der / is a trademark of
OmniScriptum GmbH & Co. KG
Heinrich-Böcking-Str. 6-8, 66121 Saarbrücken, Deutschland / Germany
Email: info@svh-verlag.de

Herstellung: siehe letzte Seite /
Printed at: see last page
ISBN: 978-3-8381-5139-7

Zugl. / Approved by: Innsbruck, Universität, Diss., 2015

UNIVERSITY OF INNSBRUCK

Abstract

Mathematics, Computer Science and Physics

Institut für Experimentalphysik

Doctor of Science

2D Arrays of Ion Traps for Large Scale Integration of Quantum Information Processors

by Muir KUMPH

Quantum computation and simulation is an emerging disruptive technology. Only first suggested by visionaries [1, 2] in the 1980s, in the last 30 years, small quantum computers have become a reality. Small systems of about ten trapped atomic ions, each mapped to a single qubit, have provided the highest fidelity quantum computations and simulations to date. If such systems were to be scaled-up, this would already give their human users immense advantages in the fields of natural simulations, search and cryptography. In order to scale-up the use of these systems, several two dimensional (2D) arrays of planar-electrode ion traps were designed, simulated, built and tested herein. The 2D arrays presented here have electronic addressability built into them. By addressable, it is meant that the control of which ion in the trap array participates in any given operation is explicit. A method to address interactions between nearest neighbors in the 2D array using an adjustable radio-frequency voltage is demonstrated by loading calcium ions into the traps and manipulating them. The theory of operation, the design methodology and the method of fabrication of the ion trap arrays is also given.

UNIVERSITY OF INNSBRUCK

Zusammenfassung

Mathematics, Computer Science and Physics
Institut für Experimentalphysik

Doctor of Science

2D Arrays of Ion Traps for Large Scale Integration of Quantum Information Processors

by Muir KUMPH

Quantenrechnen und Quantensimulationen sind eine aufkommende, bahnbrechende Technologie. In den 1980-er Jahren von Visionären vorgeschlagen [1, 2], sind einfache Quantencomputer in den letzten 30 Jahren Realität geworden. Kleine Systeme, die aus ungefähr zehn gespeicherten Ionen besteht, wobei jedes Ion ein Qubit kodiert, halten den Rekord bezüglich höchste Güte für Quantenrechnen und Quantensimulationen. Derzeit wird untersucht, ob man solche Quantensysteme erfolgreich vergrößern kann. Gelänge es, diese auf größere Quantensysteme und komplexeres Rechnen hochzuskalieren, wäre dies der Schlüssel für neue Perspektiven in den Bereichen der Naturwissenschaften, Suchalgorithmen und der Kryptographie. Mit dem Ziel der Realisierung von skalierbaren Ionenfallen wurden zweidimensionale Gitter von Ionenfallen entwickelt, simuliert und getestet. Die hier entworfenen Ionenfallen sind elektronisch aufrufbar, das heißt, dass man gezielt steuern kann, welches Ion im Gitter ansprechbar sein soll oder welche Ionen im Gitter eine Wechselwirkung haben sollen. Als Methode, um die Wechselwirkung zwischen benachbarten Ionen zu steuern, wurde eine justierbare Elektrode zwischen gespeicherte Ionen gesetzt. Im Experiment konnte deren Funktion erfolgreich demonstriert werden. Die Theorie über den Ablauf, das Versuchsdesign und die Herstellung der Ionenfalle ist ebenso beschrieben.

Acknowledgements

So many folks helped me when I asked, suggested things when I was stumped and explained things when I was wrong (sometimes patiently shaking their head while I put my foot in my mouth). To name a few: Christian, Piet, Wolfgang, Hartmut, Gerhard, Tracy, Rene, Nikos, Cornelius, Markus, Lukas, and Mike C were all instrumental in nudging me along in the right direction. I owe a considerable amount to Michael Baranov for patiently explaining quantum mechanics to me. Most supportive have been my Professors Rainer Blatt and Michael Brownnutt, who gave me the support and freedom to explore and understand the world of quantum optics. I must certainly thank Tony, Armin, and Helmut in the machine shop, who built so many of the parts of the experiment. I must thank Wolfgang and Gerhard in the electronic shops, who gave us electronics and fixed broken pieces. Without the secretaries of the Institut für Experimentalphysik, especially Patricia Moser, the logistics of starting a life in a new country and new language would have been overwhelming. I owe a special thanks to my partner-in-crime Michael Niedermayr, who I had the pleasure to build our laboratory together with. More recently, Philip, Kirsten, Adam, Kirill and Martin also worked on this project solving some very hard problems. I also owe more than I can say to Klaus, Tommy and Rene, my roommates, who showed me how beautiful life in Tirol outside the laboratory can be. Finally and foremost, I am so glad to have my wife Bernadette and son Levin as a source of inspiration. They have shown me the spark of life.

Contents

List of Figures

List of Tables

Abbreviations

2D	2 Dimensional
3D	3 Dimensional
AAO	Anodic Aluminum Oxide
AC	Alternating Current
AOM	Acoustic Optical Modulator
COM	Center-Of-Mass
CNOT	Controlled-NOT
DC	Direct Current
DDS	Direct-Ditigal Synthesis
FR4	Flame Retardant 4
ECDL	External Cavity Diode Laser
EMI	Electro-Magnetic Interference
EOM	Electro-Optical Modulator
h.c.	hermitian conjugate
ITO	Indium Tin Oxide
NF	Noise Factor
PBS	Polarizing Beam Splitter
PCB	Printed Circuit Board
PDH	Pound Drever Hall
PMT	Photo Muliplying Tube
POM	Polyoxymethylene
PSD	Power Sprectral Density
PTFE	Polytetrafluoroethylene
QECC	Quantum Error-Correcting Codes
QHO	Quantum Harmonic Oscillator
QIP	Quantum Information Processing
QIT	Quadrupole Ion Trap
RF	Radio Frequency
RF null	Radio Frequency (electric field) null

RMS	Root Mean Squared
UHV	Ultra High Vacuum

Physical Constants

Speed of Light	c	$=$	$2.997\,924\,58 \times 10^8 \ \mathrm{m \cdot s^{-1}}$ (exact)
Planck constant	$\hbar$	$=$	$1.054571726(47) \times 10^{-34} \ \mathrm{J \cdot s^{-1}}$
charge of ion	q	$=$	$1.602176565(35) \times 10^{-19} \ \mathrm{C}$
Bohr magneton	μ_{B}	$=$	$9.27400968(20) \times 10^{-24} \ \mathrm{J \cdot T^{-1}}$
atomic mass unit	u	$=$	$1.660538921(73) \times 10^{-27} \ \mathrm{kg}$
Boltzmann constant	k_{B}	$=$	$1.3806488(13) \times 10^{-23} \ \mathrm{J \cdot K^{-1}}$
vacuum permittivity	ϵ_0	$=$	$8.854187817... \times 10^{-12} \ \mathrm{F \cdot m^{-1}}$
vacuum permeability	μ_0	$=$	$4\pi \times 10^{-7} \ \mathrm{H \cdot m^{-1}}$
number pi	π	$=$	$3.141592654...$
mathematical constant	e	$=$	$2.718281828...$

physical constants from CODATA 2010 [3]

Symbols

a	distance between ions	m
a_t	distance between RF null points	m
a_n	RMS size of trapped ion's motion for quantum number n	m
a_x, a_y, a_z	trapping parameters due to static quadrupole field	dimensionless
d	ion-electrode distance	m
D	characteristic distance	m
$\boldsymbol{E}$	electric field	$\mathrm{V} \cdot \mathrm{m}^{-1}$
$\bar{\boldsymbol{F}}$	average force on ion	N
F_a	noise factor	dimensionless
$\boldsymbol{k}$	wave vector of laser light	$\mathrm{rad} \cdot \mathrm{m}^{-1}$
m_I	ion mass	kg
P	power	$\mathrm{J} \cdot \mathrm{s}^{-1}$
q	charge of ion	C
q_x, q_y, q_z	trapping parameters due to time varying quadrupole field	dimensionless
Q	quality factor	dimensionless
$\boldsymbol{r}$	position vector	m
s	saturation parameter	dimensionless
t	time	s
U	energy	J
$\boldsymbol{v}$	velocity of ion	$\mathrm{m} \cdot \mathrm{s}^{-1}$
V_{adj}	adjustable RF voltage	V
V_{nom}	main or nominal RF voltage	V
V_r	ratio of adjustable to nominal RF voltage	dimensionless
x, y, z	coordinates in space	m
x_s	position of x stretch mode of ions	m
x_c	position of x center-of-mass mode of ions	m
x_D	average displacement of ion from RF null, due to a static field	m
$\alpha_s, \beta_s, \gamma_s$	x, y and z quadrupole parameters of the static electric potential	$\mathrm{V/m}^2$

$\alpha_t, \beta_t, \gamma_t$	x, y and z quadrupole parameters of the time varying electric potential	V/m^2
$\beta_x, \beta_y, \beta_z$	characteristic trapping exponent	dimensionless
β_{FM}	FM modulation index	dimensionless
γ	natural linewidth of atomic transition	rad $\cdot$ s^{-1}
δ	detuning of laser light from transition	rad $\cdot$ s^{-1}
δ_{eff}	effective detuning of laser cooling light	rad $\cdot$ s^{-1}
$\delta\omega$	trap frequency splitting	rad $\cdot$ s^{-1}
Δ	detuning of laser from atomic transition frequency	rad $\cdot$ s^{-1}
η	Lamb-Dicke parameter	dimensionless
η_I	fraction of RF current	dimensionless
$\kappa^{(D)}$	dipole efficiency factor	dimensionless
$\kappa_x, \kappa_y, \kappa_z$	x, y and z trap efficiency factor	dimensionless
κ_d	trap-depth efficiency factor	dimensionless
κ_{dc}	doppler cooling damping coefficient	m^{-1} $\cdot$ s
μ	dipole moment	C $\cdot$ m
ν	reference oscillator frequency	rad $\cdot$ s^{-1}
ρ	state matrix	dimensionless
ρ_{ee}	excited state population	dimensionless
$\sigma_x, \sigma_y, \sigma_z$	Pauli matrices	dimensionless
ϕ	electrical potential energy	V
ϕ_{DC}	static trap voltage	V
ϕ_{RF}	time varying trap voltage	V
ϕ_s	trap pseudopotential	J or eV
Φ_{depth}	trap depth (pseudo)potential	J or eV
ω	frequency of electromagnetic radiation	rad $\cdot$ s^{-1}
ω_C	carrier (or resonant) frequency of atomic transition	rad $\cdot$ s^{-1}
$\omega_x, \omega_y, \omega_z$	secular trap frequencies	rad $\cdot$ s^{-1}
Ω_{rf}	trap drive frequency	rad $\cdot$ s^{-1}
Ω	Rabi frequency	rad $\cdot$ s^{-1}
Ω_0	effective Rabi frequency	rad $\cdot$ s^{-1}

"If the result confirms the hypothesis, then you've made a measurement. If the result is contrary to the hypothesis, then you've made a discovery." – Enrico Fermi

Chapter 1

Introduction

The impact of the information revolution on humankind cannot be overstated. The very idea that information is mutable and able to traverse physical domains is a fundamental and powerful tool. Machines which process, record and transmit information have vastly improved mankind's abilities to predict and interact with the natural world. Quantum information processing may extend such abilities greatly, and most probably in ways that cannot yet be imagined. However, the process of developing a quantum computer requires that the individual logical components of the computer are quantum in nature, individually controllable and of high fidelity. It is worth spending a little time, laying out what information is, what the limits of classical computers are, and how quantum information may change all that for the better.

Information in computers is represented by a physical state. These states are then stored and processed by the computer. One of the first applications of computers was to perform mathematical operations. Computers use physical mechanisms, such as diodes and transistors, or historically gears and levers [4] so that machines instead of humans can compute, simulate, record or transmit numbers. And in fact, any machine which computes via algorithms is understood to be equivalent to a Turing machine [5–7], where a Turing machine is a universal computer with minimal capabilities [6].

Many kinds of information can be encoded as numbers. And the information encoded into computers has grown to include music and pictures as well as the design and simulation of entire working machines. The fact that so many things can be measured, recorded and computed has fundamentally changed what people can do. The collection and computation of health statistics has allowed medicine to advance by determining which treatments actually work and which treatments only appear to work. It has allowed the sound of the human voice to be transduced into electrical signals, where it is digitized, copied, transmitted and even translated into text automatically. Networks of sensors and computers running weather algorithms allow it to be determined how storms develop, saving countless lives. Indeed, the information revolution has changed our society forever.

But for all the amazing things modern technology can do, there remain many computational problems that so-called *classical* computers will likely never solve. This statement raises two questions: What is meant by classical computers as opposed to quantum computers? And why are some problems so difficult? The first question is harder to answer than the first. Because, as it turns out, human experience is almost solely predicated on classical manifestations of the world. The world is also filled with non-classical phenomena of a *quantum* nature, but they rarely make themselves available to the human senses. The second question as to what sort of problems appear to be unsolvable with classical information technology, is easier to explain.

The rest of the this introduction is organized as follows: First, in section 1.1 the nature of pure quantum states is described and how their large number of degrees of freedom might be used to simulate or compute complex mathematics. Next, in section 1.2 the limitations of classical information processing (or more simply classical computing) will be described by giving some examples of seemingly intractable problems to simulate or compute. Finally, a few ways to perform quantum simulations, in section 1.3, and run some quantum computing algorithms, in section 1.4, are described. Specifically, a two-dimensional lattice of quantum states with nearest neighbor interactions is described and its significance to these problems is given. Finally, section 1.5 lays out how the rest of this thesis is organized.

1.1 From Classical to Quantum Computing Mechanisms

The first computing machines from such inventors as Blaise Pascal [8] and Gottfried Leibniz [9] used classical mechanical principles to make calculations. In the last few hundred years, computers have evolved into machines made from solid state electronics utilizing vast micro-miniaturization. But the basic building block of modern computing is still based on classical physical principles. In contrast, quantum information processors (QIP) use the principles of quantum physics to make calculations. Of course, the workings of modern solid-state computers can only be understood with the help of quantum mechanics. However, the states which are stored and manipulated in a classical computer have very different computational properties from those of quantum computers and simulators.

One might ask, why not use a classical computer and classical states? After all, building an apparatus to manipulate a few quantum states, might seems like the overly complex machines of Rube Goldberg [10]. While in principle, quantum states and their interactions can be modelled using classical states, it requires an exponentially increasing number of classical states to model a linearly increasing number of quantum states. The reasons for this follow from elementary quantum theory.

1.1.1 Superposition of Quantum States

Quantum states have the remarkable property that they can interfere with themselves and each other through the properties of *superposition* and *entanglement*. The canonical example of superposition is that a photon travelling from one point to another, does so as if it were a wave. And if able to travel more than one path, say through a screen with two slits, can actually travel through both slits at the same time. Figure 1.1 illustrates this classic experiment [11,

12]. A single quantum particle (on the left) encounters a wall with two slits. The particle simultaneously goes through both slits as if it were a wave. A sinusoidally varying interference pattern is observed on the photographic plate on the right, which is due to the particle interfering with itself as it lands on the screen. A simplified quantum treatment is considered next.

Photons coming from the left in figure 1.1 can be absorbed by the screen, or pass through the two slits. If the photon passes through the two slits, its wave function is split into two possible paths or quantum states. Using the bra-ket notation [13], the photon's upper path in figure 1.1 is $|0\rangle$, and the lower path is $|1\rangle$. The state of the photon's quantum state $|\psi\rangle_0$, just as it is split into the two states is

$$|\psi\rangle_0 = (|0\rangle + |1\rangle)/\sqrt{2} \tag{1.1}$$

so that the photon is said to be in a superposition of both states. Any eigenstate $|n\rangle$, such as these two different paths, undergoes an evolution of its phase as it propagates. For a photon, this is a change in its phase and is given by

$$|n\rangle\,(t) = e^{-i2\pi L/\lambda}\,|n\rangle\,(t=0), \tag{1.2}$$

where t is the time, L is the distance of the path that the photon has travelled since $t=0$, and λ is the wavelength of the photon. For the two different paths, a phase difference $\phi = \Delta L/\lambda$ proportional to the difference in the path length ΔL of the photon from each slit is accumulated. Figure 1.1 shows the geometry of the path length difference.

By the time the photon reaches the photographic plate, the photon's final state $|\psi\rangle_{\text{final}}$ is given by

$$|\psi\rangle_{\text{final}} = (e^{-i\phi_0}\,|0\rangle + e^{-i\phi_1}\,|1\rangle)/\sqrt{2} = e^{-i\phi_0}(|0\rangle + e^{-i\phi}\,|1\rangle)/\sqrt{2}. \tag{1.3}$$

The photon is still in a superposition of having gone through both slits, but now there is a phase difference ϕ which describes the different paths, which the photon has taken.

If we say that after the photon goes through either one of the two slits it is equally likely for it to land somewhere on the photographic plate with a state $|x\rangle$ (ie. $\langle x|0\rangle = 1, \langle x|1\rangle = 1$), so that the state of the photon when it hits the plate is then just the sum of the contributions from each path $|\psi\rangle = (|x\rangle + e^{-i\phi}\,|x\rangle)/\sqrt{2}$. This phase difference is what gives rise to the interference pattern on the plate, since the probability of measurement is given by $P(\Delta L) = |\langle x|\psi\rangle|^2 =$. The result is then $P(\Delta L) = 1 - \cos^2(\Delta Lk)$. The path length difference ΔL can be related to the geometry of the exact setup (slit separation and distance between screen and photographic plate).

The important features of quantum mechanics are already present in this simple example. During measurement, only the recombined photon's wave can be directly observed. And so the interference must be seen by repeating the experiment over and over till the statistics of the photon's interference can be seen. Only the relative phase difference between the two quantum states is observable and this interference is a hallmark of wave-like quantum superposition.

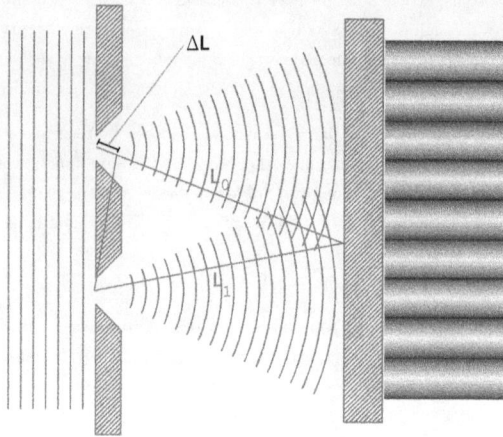

Figure 1.1: Quantum interference is much like classical wave interference, except that each individual quantum (or particle) has a probability amplitude wave associated with it. Above is a figure showing how a travelling particle (say a photon or electron) can interfere with itself by going through two slits. The particle is in a superposition of states, where the particles goes through both the top and bottom slit. The interference pattern can be seen as the probability that the particle will be observed on the screen on the right.

1.1.2 Entanglement

Somewhat more of an enigma, *entanglement* is a correlated property of quantum systems that are composed of individual particles. Entanglement is said to exist if a single quantum state composed of multiple particles **cannot be** written simply as a concatenation (or tensor product) of individual states of the particles. Discerning between a classical probabilistic mixture and entanglement can be done by checking the strength of the correlations as the measurement basis is altered. Correlations between entangled pairs of particles have long been measured [14, 15]. And this correlation is stronger, Einstein even called it *spooky* [16], than would be expected if the physics were local and realistic [17].

For example, if one has a classical probabilistic mixture of two particles, each capable of having two states, a hidden variable model can be used to represent which of the unknown two states each particle is actually in. To illustrate the idea, if there are two balls, one red and one blue. If one ball is given to Alice and the other to Bob, but packaged inside opaque boxes. Should Alice open her box and find a red ball, we know that Bob has the blue one. We can say that the hidden variable is which of the boxes the balls are in. Such a probabilistic description of quantum mechanics is not compatible with the well prepared quantum states.

In particular two quantum particles with entangled quantum mechanical states [18] cannot be represented as probabilistic mixture of particles with local hidden variables. Bell set strict limits on the correlation between pairs of classical particles with a local hidden variable, called Bell inequalities. And entangled quantum mechanical particles routinely violate these inequalities.

Quantum theory gives such correlations a degree of freedom, and these degrees of freedom apply to the combined system of particles. In other words, quantum systems composed of multiple entangled states or particles, require that one consider the whole ensemble of particles and their correlations as if they were intimately and instantly connected to each other. It is this intimate connection between entangled quantum systems that allows one to take advantage of the large computational Hilbert space and would allow for certain problems to be solved dramatically faster if a large scale quantum computer or simulator could be built.

A system which produces particles with quantum states can also produce probabilistic mixtures of particles in these states, but this is in addition to the degrees of freedom required to quantify the correlations involved in entanglement. Such mixed states are described further in this thesis, but for now only pure quantum states are considered. In particular, consider a quantum system composed of 2 photons each with a polarization state. This simple system can only be understood by considering, not only the quantum states of each photon, but also the various combinations (or correlations) of the quantum states [19, 20]. Say photon 1 and 2 have a polarization of either vertical (V) or horizontal (H). A complete set of states of the combined system of both photons is then

$$\{|v_1\rangle |v_2\rangle, |v_1\rangle |h_2\rangle, |h_1\rangle |v_2\rangle, |h_1\rangle |h_2\rangle\} \tag{1.4}$$

where each of these combinations of states can then be in a superposition. Quantum theory has predicted [20], and experiments have measured [21] these correlations. As each binary quantum state, called a *qubit*, is added, the number of combinations of the individual quantum systems doubles. So that the number of combinations is then 2^n, where n is the number of qubits in the system. Each combination has a weight compared to the other's, but it is a complex number, so that two real numbers are required to describe it. The number of classical real numbers required to describe a quantum state composed of n qubits would then be $2^{(n+1)}$. If we use the two facts that:

- the overall phase of an ensemble of qubits is arbitrary

- quantum states must be normalized, so that when measured, just one state is observed

this reduces the number of degrees of freedom by two. So that the number of degrees of freedom in a quantum system of n qubits is,

$$2^{(n+1)} - 2 \tag{1.5}$$

One can then consider, a set of n-qubits to span a Hilbert space of an exponential (in n) number of degrees of freedom.

1.2 Complexity

Many physical systems exhibit an increase in complexity as the size of the system n is increased. If the increase is either a high order polynomial ($n^p, p \geq 4$) or stronger than polynomial (for example exponential), the problem becomes intractable as it is scaled-up. Because multi-body quantum systems exhibit an exponential increase in their state space, this can be used as a powerful computational resource [2]. Below are given two examples of how complexity can increase. In the first example, an exponential increase is shown in the game of chess (1.2.1). While there is no known quantum algorithm for chess, its increase in complexity mirrors the complexity of quantum systems. Fluid dynamics complexity increases in a polynomial way as it is scaled up (1.2.2), but so far, high Reynolds number flows have remained out of reach for classical computers.

1.2.1   Chess

Perhaps, the most accessible example of classically intractable problems comes from games. For instance, the game of chess. Some might say, that the game of chess is solved, since in 1998, the world's best chess computer, IBM's Deep Blue, beat the world's best human player, Gary Kasparov. However, the game is not solved. The difference is considerable.

What the world's best chess computer does is run a clever and complex algorithm, written by a massive team of expert chess programmers. The speed of the computer allows the chess heuristics of these experts to be performed on many possible chess moves. The algorithm then selects the move with the highest score according to the heuristics. What the computer cannot do is guarantee a win or a tie against a theoretically perfect player. It may be that there are, as yet, unknown tactics or strategies, which Kasparov did not know, which would stump a computer like Deep Blue.

While chess may seem like a trite example, it turns out that there are many problems, which like chess, explode in complexity. The reason for the difficulty (at least in chess) has been well analysed [22]. To actually solve chess, one would have to check every possible move playable by both players. There are simply too many possible moves to check. For example, in the beginning of the game, there are 20 possible first moves. So after the first two players take a turn, there are $20^2 = 400$ possible board positions. As an approximation, each player has about 30 possible valid moves each turn. The number of possible chess games then grows exponentially, so that after each player takes n turns, the number of possible games is A^{2n}, where A is of order 30. Since an average game has about 40 moves (80 turns), this means that the number of possible chess games is of order 10^{120}. The number of observable atoms in the universe is estimated to be 10^{80}. If we were able to encode the outcome of all the possible games (white wins, black wins, or tie) onto just one atom, there are not enough atoms in the universe to do so. By an astounding factor of roughly 10^{30}. If a computer were to sequentially check each possible game outcome at a rate of one billion per second, it would take 10^{87} years [22] to calculate what the first move(s) should be to ensure a win or tie. This may seem like a pedantic exercise, but there are many problems in nature which grow so quickly, that like chess, no direct computation of the solution via classical computation is possible. There are certainly computer projects, like Deep-Blue, where a seemingly good solution exists. But there is no knowledge of whether such a solution is optimal, or even how close to optimal it is. If one were to use a quantum Hilbert space to encode the result of each possible game of chess onto a quantum state, it would require approximately

$\log_2(10^{120})/2$ or approximately 180 qubits. There is unfortunately no known quantum algorithm to allow for a solution for chess to be computed, but the vast size of the Hilbert space illustrates the possible power of quantum information processing.

1.2.2 Fluid Dynamics

Perhaps a more practical example exists in the area of computational fluid dynamics. A proper treatment of a viscous fluid involves solving the Navier-Stokes equations [23]. But because of the non-linearity involved, these equations can only be solved piece-wise in space and time. This means that solving for the dynamics of a turbulent boundary layer in a given system, requires computing resources which scale strongly with the Reynolds number (Re). For relevant geometries, such as one infinite dimension and 2 periodic boundary conditions, the memory requirements scale as Re^3 while the computational cost scales as Re^4 [24]. Because of this, it is often said that computational fluid dynamics requires the answer (in the form of the turbulent boundary layer dynamics) before it can compute the solution of the entire flow field.

The largest direct simulation of the Navier-Stokes equations has been of the flow between two infinite parallel sheets with a Reynolds number of 5200 [24]. It was performed on a petascale supercomputer. The disciplines of naval architecture and meteorology have Reynolds numbers in the range of 10^4 to 10^{10} [25]. If indeed, "The Free Lunch is Over" [26] and classical computers can now only get bigger (not faster), then one can use parallel processing techniques. This means one would require a machine between 10^4 and 10^{28} times larger and/or faster than the petascale effort to directly solve relevant fluid dynamics problems.

In the last 100 years, computers have increased in speed and size by roughly a factor of 10^{15}. It seems optimistic to think one would be able to scale existing computer hardware up by such a large amount. It is likely then that the solution of turbulent fluid flow for a Reynolds number of 10^{10} will likely never be directly solved by current computer architecture.

1.3   Quantum Simulations

The amount of computational resources required to directly simulate quantum systems scales exponentially, because its Hilbert space grows exponentially. Feynman first suggested simulating one quantum system with another quantum system [1]. Such a quantum simulator holds the promise of an exponential speed-up over trying to simulate some quantum system with a classical computer [27].

Designing a correspondence between a controllable quantum system to an interesting class of physical quantum systems is currently an active area of research [28]. For example, one active area of research is systems of interacting magnetic spins [29]. Such *Ising* models provide rich dynamics and can provide insight into condensed matter phenomena, such as high temperature superconductivity [30]. Figure 1.2 shows a conceptual graph of a two-dimensional (2D) lattice of quantum mechanical spins. Each spin is located on the circular nodes of the lattice. The interactions are shown by the lines between the nodes. Since magnetic spins are two-level quantum

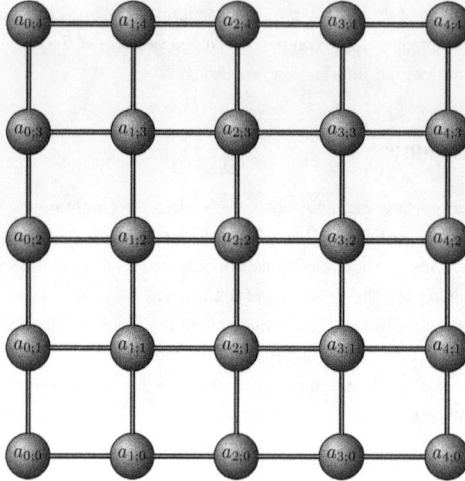

Figure 1.2: Depiction of a two-dimensional square lattice of interacting magnetic spins (Ising model [29]). Each spin is represented by a circular node. The interactions are represented by the lines between the nodes. These sorts of systems could give insight into condensed matter systems of interest, such as high-temperature superconductivity. The spins can be modelled by qubits. For highly entangled systems, exceeding approximately 30 qubits, classical computation becomes intractable [30].

systems, such a lattice of spins could be simulated with a 2D lattice of two-level quantum systems (also called *qubits*), each with nearest neighbor interactions. Building controllable systems capable of simulating this kind of physics [31] would allow for the above mentioned phenomena to be better understood.

1.4 Quantum Computation

Deutsch realized that a quantum system could also be used to build a Universal Quantum Computer [2]. Many real-world problems can be solved faster either by increasing the number of computational cycles per unit time, or by increasing the size of the data processed in a single cycle. There are some limitations in quantum computation, because the vast Hilbert space of quantum systems cannot be directly accessed, as the process of measurement alters the system. Nonetheless, utilizing this vast quantum Hilbert space, some algorithms achieve an exponential speed-up in computation. Below two quantum computing algorithms, which show an exponential increase in speed over the best known classical computing algorithms are described.

1.4.1 Quantum Linear Equation Solver

An algorithm using a quantum computer which would allow for an exponential speed-up in the approximation to a system of linear equations has been developed [32] and tested [33].

Since all dynamical systems, from economics to fluid dynamics and weather models, can be linearized with sufficiently fine sampling, this could considerably improve the ability to solve complex systems, such as fluid-dynamics, and predict their behavior.

1.4.2 Shor's Algorithm

In the area of number theory, Shor's algorithm [34] has demonstrated that the Hilbert space of a system of qubits can be used to factor numbers in polynomial time with respect to the size of the number. This would have huge implications for military and economic security, since much of modern cryptology relies on the difficulty of factoring large numbers.

1.5 Towards Quantum Computing and Simulation

It is not meant to be claimed here that all problems, which grow exponentially in classical computation (like chess) can be solved more efficiently with the use of quantum computers. But there exist certain classes of problems (such as number factorization and linear equation approximation) which scale unfavorably with known classical algorithms, yet would be very efficiently solved with the use of a quantum computer. The theoretical work on quantum algorithms is still quite recent, and it may be that great advances will be made in the near future. However it is hard to motivate people to work on quantum algorithms when no such computer yet exists to run them. The rest of this thesis is concerned with the physical architecture of a machine, which could run such algorithms and simulations in an economically useful manner. Specifically, this thesis describes a machine for manipulating the quantum states of trapped atomic ions with sufficient fidelity, that information, when encoded in these states, could be processed in a way to perform complex computations and simulations.

The rest of this thesis is organized as follows. Chapter 2 outlines the a-priori state-of-the-art techniques and theoretical foundations in using trapped atomic ions for quantum information processing (QIP). Chapter 3 describes how to scale-up these techniques using an addressable two-dimensional (2D) array of trapped atomic ions [35]. The next chapter describes the technical details of the laser sources and vacuum chamber for testing 2D arrays of trapped $^{40}Ca^+$ ions. The next three chapters describe three increasingly miniaturized charged particle trap arrays. Chapter 5 details the design, simulations and testing of a dust trap called "Dusty". Chapter 6 gives the design, simulations and testing results of an RF addressable ion trap called "Folsom". In chapter 7 microscopic ion trap arrays currently being testing and fabricated are described. Lastly, chapter 8 looks forward to explore how 2D arrays of trapped ions might contribute towards large scale quantum simulations and computation and what key technologies need to be developed towards this aim.

Chapter 2

Experimental Techniques for QIP with Ions

There are many physical systems which can be used to store and manipulate quantum states. Researchers in the field of experimental quantum physics are now able to perform rudimentary quantum computing and simulation using various physical systems, such as photons [36], circuit-qed [37], quantum spintronics [38], atoms [39] and ions [40]. For many purposes, trapped atomic ions represent a nearly ideal system for storing and manipulating quantum states. The ion's bound electrons can be used to store information in *electronic* states and the motion of the ion in the trap can be used to store information in *motional* states. Moreover, the motion of one trapped ion can be coupled to other trapped ions.

Because a trapped ion is a vibrating charge, it can interact resonantly with other trapped charged particles via the Coulomb force with the mathematical form of a dipole-dipole interaction. The magnitude of the dipole moment of trapped ion is $|\mu| = qa_n$, where q is the charge of the ion and a_n is the extent of the ion's motion in the trap. For many experiments with trapped $^{40}\text{Ca}^+$ this distance is of order 10 nm.

An atomic ion also has bound electrons, with electronic states. These electrons have a motional extent of roughly $1\,\text{Å}\,(0.1\,\text{nm})$. So that the dipole moment of the ion's motion is of order 100 times larger than the dipole moment of the electrons' motion bound to the ion. Because of this large difference, these states can be used for different purposes.

Considering resonant dipole-dipole coupling between two trapped ions. The rate of such coupling then goes as the dipole moment of each multiplied against each other. For a given separation distance, any interaction between the ions' motional states is then 10^4 faster than any interaction between the ions' electronic states. This means that two trapped ions can be trapped, so that their motional states are resonantly coupled, but their electronic states are well isolated. For this reason, the motional states of the ions are most often used to perform interactions between the ions [40], forming a kind of information *bus*. Whereas, the electronic states are well isolated and can be used to form a quantum memory.

Trapped ions, with their electronic and motional states and dipole couplings, are not the only ingredients for quantum information processing with ions. Lasers are a key tool. Using focussed beams of laser light directed at the ions, the electronic states can be manipulated, or an interaction between the motional states of a trapped ion and its electronic states can be produced. This allows the motional and electronic states of trapped ions to be manipulated, stored and transmitted.

This chapter gives a broad overview of the pertinent state-of-the-art techniques used to trap ions and manipulate their motional and electronic quantum states with (mainly) $^{40}Ca^+$ ions. The theoretical quantum mechanics and atomic physics, which these techniques are based upon, are also reviewed. Many of the techniques described are used to produce the experimental results of trapping and manipulating trapped $^{40}Ca^+$ ions given in chapter 6. Moreover, these experimental techniques, with careful design, could be scaled-up to allow large scale quantum information processing. In chapter 3, a scalable architecture using 2D arrays of addressable ion traps is described. This architecture is designed to primarily use the experimental techniques described here in this chapter.

The rest of this chapter is organized as follows. First, a classical description of radio-frequency (RF) quadrupole ion traps is given in section 2.1. The pertinent features of the ion's electronic states are given in section 2.2. Section 2.3 gives the theory of the trapped ion's quantized motion. The theory of the interaction of laser light with the ion's motional and electronic states is given in section 2.4. Photoionization loading of the trap is described in section 2.5. Section 2.6 then describes the process of laser cooling, including sideband cooling which allows for the ion's motional state to be initialized. Qubit rotations and gates are then introduced in section 2.7 and the method of qubit measurement is given in section 2.8. The error syndromes of qubit decay and decoherence are explained in section 2.9. As heating is a particularly pathological source of decoherence for ion traps, the methods to measure heating rates are described in section 2.10. The ions trapped using the techniques described here exhibit a kind of oscillating motion, which is usually desirable to minimize, called micromotion. This micromotion and methods to minimize it are described in section 2.11. Finally, a few architecture proposals for large-scale trapped ion computation are considered in section 2.12.

2.1 Trap Basics

In typical experiments involving atomic ions, a positively charged atomic ion (cation) is produced by removing an electron from a neutral atom. This positively charged ion responds to electric and magnetic fields and can be trapped by various means such as Penning [41], quadrupole [42], and Kingdon [43] ion traps. For the work presented here, quadrupole ion traps (QIT), which use oscillating radio-frequency (RF) electric fields, were used. A positively-charged ion then responds to these electric fields, so that it can be held at a fixed point in space (typically under vacuum) enabling experiments to be performed on it. In what follows, starting from basic electrostatics (2.1.1), the classical (non-quantized) physics of the quadrupole ion trap is explained (2.1.2). The approximation of this trap to a harmonic trap (2.1.3) is then described.

An introduction to the electrode structure, which can provide the necessary trapping fields, is given (2.1.4). Lastly, the energy depth of the trap is described using the trap depth (2.1.5).

2.1.1   Electrostatics

Consider an ion at position $\boldsymbol{r}$ with a charge, q, in the presence of an electric field, $\boldsymbol{E}(\boldsymbol{r})$, at an instant in time. It experiences a force, $\boldsymbol{F} = \boldsymbol{E}(\boldsymbol{r})q$. Maxwell's laws dictate that in the absence of time varying magnetic fields, the electric field $\boldsymbol{E}$ is related to the gradient of the field's potential ϕ as $\boldsymbol{E} = -\boldsymbol{\nabla}\phi$. The electrical potential ϕ is then related to the charge density ρ in space by $\boldsymbol{\nabla}^2\phi = \rho/\epsilon_0$, where ϵ_0 is the vacuum permittivity. While electrodes are used to create the electric field which traps the ion, it is necessary to hold the ion in a vacuum away from other materials, so as to isolate it and keep it from reacting with other atoms. In an ideal vacuum, the space charge is zero, and so the electrical potential follows Laplace's equation,

$$\boldsymbol{\nabla}^2\phi = 0, \text{ so that, } \boldsymbol{\nabla} \cdot (\boldsymbol{\nabla}\phi) = -\boldsymbol{\nabla} \cdot \boldsymbol{E} = 0. \tag{2.1}$$

This differential equation can be solved everywhere in space, so long as the boundary conditions of the potential are specified. Using Cartesian coordinates, $\boldsymbol{\nabla} = \{\partial/\partial x, \partial/\partial y, \partial/\partial z\}$, the divergence of the electric field then follows the relation

$$\boldsymbol{\nabla} \cdot \boldsymbol{E} = \frac{\partial E_x}{\partial x} + \frac{\partial E_y}{\partial y} + \frac{\partial E_z}{\partial z} = 0. \tag{2.2}$$

The gradients of the electric field in each direction are themselves scalar fields. Considering the static (in time) part of the electric field, these scalar gradients can be described by the parameters $\alpha_\mathrm{s}, \beta_\mathrm{s}$ and γ_s. The sign convention is chosen here so that, for each Cartesian coordinate, a positively charged particle can be trapped in a potential minimum when the respective parameter is positive. These parameters are then,

$$\alpha_\mathrm{s} = -\frac{\partial E_x}{\partial x}, \beta_\mathrm{s} = -\frac{\partial E_y}{\partial y}, \gamma_\mathrm{s} = -\frac{\partial E_z}{\partial z}. \tag{2.3}$$

Using the divergence relation of equation 2.2, the gradients of the electric field at each point in space are constrained as follows,

$$\gamma_\mathrm{s} = -(\alpha_\mathrm{s} + \beta_\mathrm{s}). \tag{2.4}$$

If the electric field is zero at the origin and $\alpha_\mathrm{s}, \beta_\mathrm{s}$ and γ_s are constant, then equation 2.2 can be integrated to give the electric field as,

$$\boldsymbol{E}(x, y, z) = -(\alpha_\mathrm{s}x\hat{\mathbf{i}} + \beta_\mathrm{s}y\hat{\mathbf{j}} + \gamma_\mathrm{s}z\hat{\mathbf{k}}), \tag{2.5}$$

where $\hat{\mathbf{i}}$, $\hat{\mathbf{j}}$, and $\hat{\mathbf{k}}$ are unit vectors in the directions x, y, and z respectively. Regions in space around an electric-field null point, where $\alpha_\mathrm{s}, \beta_\mathrm{s}$ and γ_s are constant is mathematically equivalent to considering just the second order spherical harmonics (called the *quadrupole* terms) in an internal multipole expansion of the potential [44, 45]. Building electrode structures and operating them so that just the quadrupole terms need be considered is explained later in this thesis.

The electrical potential of a field is then simply the path integral of the electric field. So that,

$$\phi_B - \phi_A = \int_A^B -\boldsymbol{E} \cdot \boldsymbol{dl} \qquad (2.6)$$

where $\boldsymbol{dl}$ is the differential vector on the path from point A to point B. If the electric field is of a quadrupole form (eq. 2.5) and the origin is defined to have zero potential, then integration yields a potential of

$$\phi(x, y, z) = \frac{\alpha_s x^2 + \beta_s y^2 + \gamma_s z^2}{2}. \qquad (2.7)$$

Since equation 2.4 requires that the sign of at least one of α_s, β_s, and γ_s be of opposite sign from the others, it follows that any electric field can form a potential minimum in at most two dimensions. In fact, Earnshaw's theorem [46] states that no point charge or collection of them is stable under the influence of electrostatic forces in free space. The potential energy of a particle U with a charge q is given by

$$U = q\phi. \qquad (2.8)$$

Consequently, if a potential minimum is formed around some point in one or two direction(s) in space, in at least one other dimension, any physical electric field will try to eject a charged particle from this location. Such a point is called a *saddle point* in the potential. At a saddle point of the potential, the electric field is zero in all directions, and is trapping in one or two directions, while it is anti-trapping in the other directions. While this is true for any instant in time, time-varying electric fields, interacting with a charged particle with a mass m, can give rise to a time averaged force which has a trapping behavior in all three dimensions.

Such a trap based upon time-varying fields is called a quadrupole ion trap [42, 47–49], and it traps a charged particle at the electric-field null (or saddle point) by using time varying and (optionally) static quadrupole electric fields. The time dependent potential of such a trap is given by

$$\phi(x, y, z, t) = (\alpha_s x^2 + \beta_s y^2 + \gamma_s z^2)/2 + \cos(\Omega_{rf} t)(\alpha_t x^2 + \beta_t y^2 + \gamma_t z^2)/2, \qquad (2.9)$$

where t is the time, Ω_{rf} is the angular frequency of the oscillating component of the electric field, and α_s refers to the gradient in the x-direction of the static electric field and α_t refers to the gradient in the x-direction of the time dependent part of the electric field. Similarly named, are the static and time dependent parts of the y and z direction gradients of the electric field.

There are two basic classes of quadrupole ion traps (QIT), both of which are constrained by equation 2.4 One type is called a linear QIT, where one of the time varying gradients is set to zero; for instance $\alpha_t > 0$, $\beta_t = -\alpha_t$, $\gamma_t = 0$. This then creates a time averaged force which has a trapping behavior in only two dimensions. A static electric field can then be used to confine the charge in the z direction. In this thesis we are concerned with another type of trap which has $\alpha_t > 0$, $\beta_t > 0$, $\gamma_t = -\alpha_t - \beta_t < 0$, so that a time averaged force creates a trapping behavior in all three dimensions (3D). Such a trap is called a *point* or 3D quadrupole ion trap (QIT). In the following section, it is described how this electric field can give rise to a time averaged trapping force.

2.1.2 Mathieu Equation

Restricting initial consideration to the motion along the x-axis, as the other directions y and z are uncoupled, the force on a particle in the x direction is given by

$$F_x = E_x q = m\frac{d^2x}{dt^2}, \tag{2.10}$$

where the electric field is an oscillating quadrupole of form 2.9 so that

$$E_x(t, x) = -x(\alpha_t \cos(\Omega_{\mathrm{rf}}t) + \alpha_s). \tag{2.11}$$

The equation of motion of the ion can then be written as,

$$m\frac{d^2x}{dt^2} = -qx[\alpha_t \cos(\Omega_{\mathrm{rf}}t) + \alpha_s]. \tag{2.12}$$

By using the dimensionless parameter $\xi = \Omega_{\mathrm{rf}}t/2$, the equation can be rewritten in the standard form of the Mathieu differential equation [50, 51]

$$\frac{d^2x}{d\xi^2} + [a_x - 2q_x \cos(2\xi)]x = 0, \tag{2.13}$$

where

$$a_x = \frac{4q\alpha_s}{m\Omega_{\mathrm{rf}}^2}, \tag{2.14}$$

$$q_x = -\frac{2q\alpha_t}{m\Omega_{\mathrm{rf}}^2} \tag{2.15}$$

are called the trapping parameters. For a quadrupole ion trap (QIT), which traps ions in all three dimensions using the time varying potential, the oscillating parts of the electric field are constrained by the equation 2.4, so that $\gamma_t = -(\alpha_t + \beta_t)$. Traps, which trap in all three dimensions using the time varying potential, but do not have cylindrical symmetry (i.e. $\alpha \neq \beta$), are sometimes called "elliptical" traps [52]. Here we consider a 3D trap, where the gradient in the x and y directions are equal ($\alpha_t = \beta_t, \alpha_s = \beta_s$). So that using the definition of the trapping parameters (eq. 2.14), the trapping parameters in each direction are related as

$$q_z = -2q_x, a_z = -2a_x. \tag{2.16}$$

In the absence of other fields, each coordinate axis of the quadrupole ion trap is associated with an uncoupled Mathieu equation. The Mathieu equation is a special form of the Hill equation [53, 54], so that the Floquet theorem [55] applies. The Floquet theorem gives that such an equation can be solved with an ansatz of the form,

$$u(\xi) = e^{\mu_x \xi}\phi(\xi) = e^{\mu_x \xi}\sum_{r=-\infty}^{\infty} C_{2r}e^{2ri\xi} = \sum_{r=-\infty}^{\infty} C_{2r}e^{(\mu_x + 2ri)\xi}, \tag{2.17}$$

where the coefficients C_{2r} decrease as r increases for stable solutions; i.e. for small q_x, $C_0 = 1$ and $C_2 = q_x/4$. For stable solutions, $\mu_x = i\beta_x$ is purely imaginary, so that β_x is real and called the characteristic exponent in the x-direction. The solution $u(\xi)$ with this ansatz is complex and so the real solution of the motion is given as a linear combination of $u(\xi)$ and its complex conjugate. Substituting the time variable t back in for ξ, the solution to the Mathieu equation is,

$$x(t) = Au(t) + Bu^*(t). \tag{2.18}$$

Where for an even solution (cos), $A = B$. Since the slowest motional frequency of this system is given by $e^{i\beta_x \xi}$ and $\xi = \Omega_{\rm rf}t/2$, this frequency can be defined,

$$\omega_{\rm x} = \beta_x \Omega_{\rm rf}/2. \tag{2.19}$$

This trap frequency $\omega_{\rm x}$ is also called the *secular* (meaning slow) frequency. By substituting the ansatz (eq. 2.17) into the Mathieu equations, a recursive set of linear equations is produced. For a solution to exist the determinant of this set of equations needs to be equal to zero [56]. The characteristic exponent β_x can then be found [51] using this condition.

Stable solutions (real values of β_x) are found for only certain ranges of the trapping parameters a_x and q_x. For a cylindrically symmetric trap, where $q_x = q_y, a_x = a_y$ and $q_z = -2q_x, a_z = -2a_x$ (see equation 2.16), the first stability region is plotted in figure 2.1 as a function of β_z and a_z. Inside this first stability region, the characteristic exponent is constrained so that $0 < \beta_z < 1$. Unfortunately, while the form of the Mathieu equation is known, the characteristic exponent does not have a closed form solution. Approximations have been developed which give the trapping parameter a_x as a function of the trapping parameter q_x and the characteristic exponent β_x in the limit of $|a_x|, q_x \ll 1$ [57, 58]. By inverting this approximation, the characteristic exponent β_x can be approximated to second order in q_x as [59],

$$\beta_x \simeq \left[a_x - \frac{a_x - 1}{2(a_x - 1)^2 - q_x^2}q_x^2 - \frac{5a_x + 7}{32(a_x - 1)^3(a_x - 4)}q_x^4 \right]^{1/2}. \tag{2.20}$$

When the trapping parameter $q_x \ll 1$, then it makes sense to approximate the characteristic exponent with just the lowest order terms of these parameters so that

$$\beta_x \simeq \sqrt{a_x + q_x^2/2}. \tag{2.21}$$

The secular frequency $\omega_{\rm x}$ in the low q_x limit is then given by,

$$\omega_{\rm x} = \frac{\Omega_{\rm rf}}{2}\sqrt{a_x + q_x^2/2}, \tag{2.22}$$

where equations 2.19 and 2.21 have been used.

2.1.3 Pseudopotential Approximation

Consider a situation when $a_x = 0$, so that the harmonic secular motion is trapped only by the oscillating field at frequency $\Omega_{\rm rf}$. The secular frequency when $q_x \ll 1$ is then

$$\omega_{\rm x} = \frac{|q_x|\Omega_{\rm rf}}{\sqrt{8}} = \frac{2q|\alpha_t|}{\sqrt{8}m\Omega_{\rm rf}}. \tag{2.23}$$

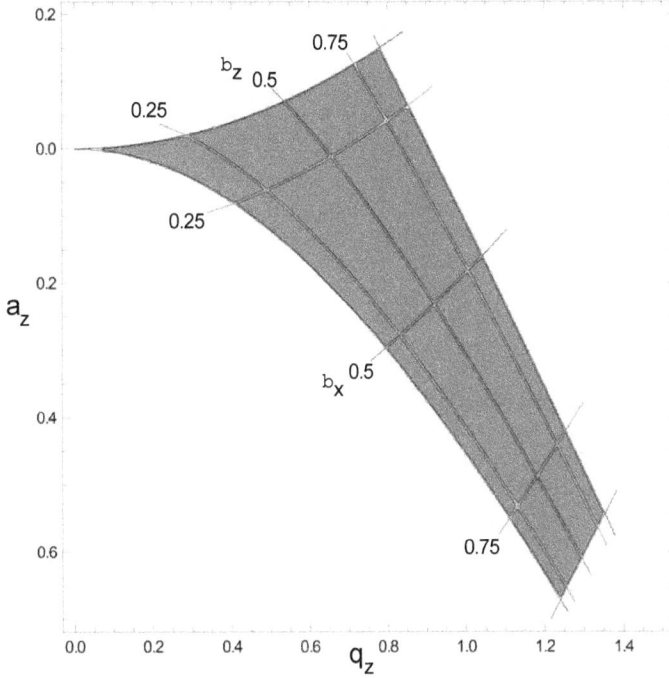

Figure 2.1: First stability region of the Mathieu equations for a spherical ion trap with respect to the trapping parameters q_z and a_z. Inside the shaded area, the Mathieu equations have a stable solution. There are an infinite number of much smaller stability regions, but most spherical ion trap experiments are performed inside this shaded region. For trap parameters outside a stable region, the ion's motion grows without bounds.

Using Hooke's law to describe a *pseudopotential* [60] associated with this harmonic motion, the pseudopotential ϕ_s is,

$$\phi_s(x) = \frac{1}{2}m\omega_x^2 x^2 = \frac{mq_x^2\Omega_{rf}^2 x^2}{16} = \frac{q^2\alpha_1^2 x^2}{4m\Omega_{rf}^2} = \frac{q^2|\boldsymbol{E}(x, y=0, z=0)|^2}{4m\Omega_{rf}^2}, \qquad (2.24)$$

where $|\boldsymbol{E}|$ is the amplitude of the time varying electric field. Treating this pseudopotential as if it were an actual potential is called the pseudopotential approximation. The potential associated with the static part of the trapping field, given by parameter a_x, can then be added separately (reproducing equation 2.22 for the secular frequency). In general, the pseudopotential field can be described as a function of position $\boldsymbol{r}$ in an electric field $\boldsymbol{E}(\boldsymbol{r})$ as,

$$\phi_s(\boldsymbol{r}) = \frac{q^2|\boldsymbol{E}(\boldsymbol{r})|^2}{4m\Omega_{rf}^2}. \qquad (2.25)$$

2.1.4 Trap Electrode Geometry

Until now, the quadrupole field has been assumed to exist. Here the shape of metallic electrodes is described, which will produce a rotationally symmetric quadrupole field. Such a trap is called a *Paul Trap*. For a cylindrically symmetric trapping potential, the trapping frequencies in the x and y direction are equal. This means that the time-varying) electric-field gradients and trapping parameters for the x and y direction are also equal, so that $\alpha_t = \beta_t$. Accordingly, equation 2.4 gives that the gradient in the z direction is given by $\gamma_t = -2\alpha_t$. The surfaces of constant potential are given by the form for a quadrupole potential in equation 2.5. Substituting the relationship between α_t, β_t and γ_t, this then gives surfaces of equipotential ϕ as,

$$\phi = \alpha_t \frac{r^2 - z^2}{2}, \tag{2.26}$$

where $r^2 = x^2 + y^2$ is the radial coordinate in the cylindrically symmetric trap. Drawn in figure 2.2 is one possible solution to equation 2.26. Equipotential lines satisfying this relation form a hyperboloid of revolution around the z-axis. The electrode shape is such that quadrupole field lines are normal to the equipotential surface. Such equipotential surfaces can be made of metal and produce a quadrupole field. The voltage potential on the electrodes is then given by

$$\phi_a = -\frac{\alpha_t z_0^2}{2}, \tag{2.27}$$

$$\phi_r = \frac{\alpha_t r_0^2}{2}, \tag{2.28}$$

where ϕ_a is the axial electrode's voltage, ϕ_r is the radial electrode's voltage, and z_0 and r_0 are the distances from the center of the trap to the axial electrode and radial electrode respectively.

The above equations relating the shape of the electrodes, the electrode voltages and the time-varying electric-field gradient $-\alpha_t$ are true for either the static component or the time varying component of the electrodes' voltage and the trap's electric field. This allows one to compute the static and time varying gradients α_s and α_t as a function of the distance of the hyperbolic electrodes from the center of the trap and the electrode's voltages. For a trap, where the axial electrodes and the radial electrodes both have the same distance d from the center of the trap, the time-varying gradient in the x direction for an ideal spherical trap, $\alpha_{I,t}$ is then

$$\alpha_{I,t} = \frac{\Delta\phi}{d^2} \tag{2.29}$$

where $\Delta\phi$ is the difference in the time-varying voltage between the radial electrode and the axial electrode. Because there are no other electrodes in the model, one is free to add a voltage to all electrodes without changing the electric field. Usually, one electrode is defined to be at zero potential so that the required static voltage ϕ_{DC} and time varying voltage amplitude ϕ_{RF} are given by

$$\phi_{DC} = \alpha_s d^2 \tag{2.30}$$

$$\phi_{RF} = \alpha_t d^2. \tag{2.31}$$

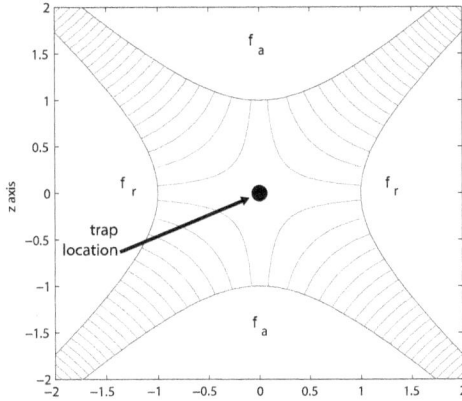

Figure 2.2: Geometry of hyperbolic electrodes. Equipotential lines satisfying the relation $\phi = \alpha_t (r^2 - z^2)/2$ form a hyperboloid of revolution around the z-axis. ϕ_a is the axial electrode voltage and ϕ_r is the radial electrode voltage. Both sets of electrodes here are drawn without units a distance $d = 1$ from the center of the trap. Actual dimensions and voltages would define the field gradient α_t. Red lines show the quadrupole field lines, which meet the trap electrodes normal to the surface. At the center of the trap, the electric field is zero and is where ions could be trapped. Equipotential surfaces with the applied voltages can be made of metal and provide the trapping field for an RF Paul trap.

Note that if one electrode is set to be at zero potential, then the center of the trap, while still having no electric field, would no longer be at zero potential. The potential at the center of the trap would then be equal to half the applied voltage.

2.1.4.1 Planar-Electrode Traps

Building traps with hyperbolic electrodes has the advantage that they produce a perfect quadrupole field. However, there are many practical drawbacks to such an electrode geometry. Such three-dimensional (3D) structures are hard to produce, and the optical access necessary for experiments is made difficult by the large solid angle of the electrodes. These drawbacks become more pronounced as the dimensions are reduced. Many proposals call for making electrodes of microscopic dimensions. The field of metal lithography, used in micro-electronics allows for accurate complex two-dimensional (2D) structures to be produced. It turns out that one need not build exact hyperbolic electrode structures as many other electrode geometries sufficiently approximate a quadrupole field near the RF electric-field null-point (called the RF null from here on). Consider the electric field, at an instant in time, produced by some arbitrary electrode geometry. Expanding the field in the x-direction, one obtains a polynomial of the form

$$\boldsymbol{E}(x) = \boldsymbol{A}_0 + \boldsymbol{A}_1 x + \boldsymbol{A}_2/2 + x^2 + ... \qquad (2.32)$$

Other directions y, z can be treated similarly. Specifically, expanding around a null point in the field ($A_0 = 0$), the first order term (A_1) in the expansion of the electric field is related to the quadrupole coefficient α_t as $A_1 = -\alpha_t \hat{\imath}$ (eq. 2.3). One can define an trap efficiency factor, κ_x, which relates the x-direction gradient at the center of the trap $-\alpha_t$, with the gradient of an ideal hyperbolic-electrode trap $-\alpha_{I,t}$ with the same applied voltage ϕ_{RF},

$$\kappa_x = \frac{\alpha_t}{\alpha_{I,t}} = \frac{\alpha_t d^2}{\phi_{RF}}, \tag{2.33}$$

where d is the distance of the center of the trap to the closest electrode. If the ions are constrained to be close to the center of the trap, so that the higher order terms in equation 2.32 are small compared to the first order terms, the non-ideal trap can be treated as an RF quadrupole trap, albeit with a geometry-dependent trap efficiency factor κ_x.

In the absence of DC fields, the abstraction of the trap electrode geometry to a trap efficiency factor κ_x allows one to write the secular frequency ω_x in terms of the applied RF voltage ϕ_{RF} as,

$$\omega_x = \frac{q\kappa_x \phi_{RF}}{\sqrt{2}d^2 m \Omega_{rf}}. \tag{2.34}$$

For a rotationally-symmetric planar-electrode geometry, it then follows that $\omega_y = \omega_x$ and $\omega_z = 2\omega_x$. In this special case, the efficiency factors are equal for each Cartesian directions x, y and z, so that $\kappa_x = \kappa_y = \kappa_z$.

In order to visualize the field lines of such a trap qualitatively, figure 2.3 shows the 2D simulation results[1] of a planar electrode QIT. Four electrodes, two with a positive voltage, and two with a negative voltage, create a quadrupole near the ion's trapping site. One of the negative electrodes is 7 mm above the trap surface created by the rest of the electrodes. The ions are trapped about 0.5 mm above the surface of this trap.

2.1.5 Trap Depth

An important characteristic of ion traps, is their ability to hold energetic ions. This ability is quantified by the trap depth ϕ_D. In the pseudopotential approximation, this is the energy an ion needs to obtain before it can escape the trap or impact an electrode. For ideal hyperbolic electrode traps, the pseudopotential increases without limit between the unending electrodes. However, the pseudopotential (eq. 2.24) just above the shortest distance d to one of the electrodes can be found using the magnitude of the electric field there. For a Paul trap, the minimum trap depth is in the radial directions x and y. An ion with enough energy to reach this point could collide with the electrodes and be lost. Using the result above for $\alpha_{I,t}$ (eq. 2.29), the trap depth is then defined for an ideal Paul trap ϕ_{DS} as,

$$\phi_{DS} = \frac{q^2 \phi_{RF}^2}{md^2 \Omega_{rf}^2}. \tag{2.35}$$

---

[1] COMSOL Multiphysics v4.3

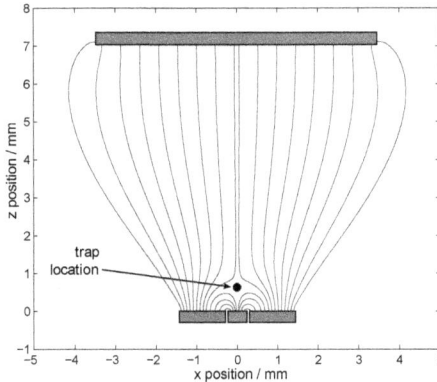

Figure 2.3: The 2D simulation results of a planar electrode quadrupole ion trap. Four electrodes, two with a positive voltage, and two with a negative voltage, create a quadrupole near the ion's trapping site. One of the negative electrodes is 7 mm above the trap surface created by the rest of the electrodes. The ions are trapped about 0.5 mm above the surface of this trap.

Traps with finite electrodes have saddle points of the pseudopotential. If an ion crosses this point, it will be accelerated further away from the trap center by the pseudopotential. The trap-depth efficiency factor κ_d is the ratio of the trap depth of any given trap, with an ion-electrode separation distance d to that of the ideal hyperbolic-electrode trap and is

$$\kappa_d = \frac{\phi_D}{\phi_{DS}}. \tag{2.36}$$

To illustrate the saddle points of the pseudopotential, an axisymmetric ring-trap geometry is depicted and simulated in figure 2.4. In figure 2.4a, the geometry of the ring trap is shown. The trap is composed of a ring of 200 μm wire diameter, with an internal diameter of 1 mm, and two spherical endcaps of diameter 200 μm separated by 1 mm. The ion-electrode distance is 500 μm. A plane shows a cross-section through the axis of the trap, in which the pseudopotential is simulated. In figure 2.4b, the pseudopotential field simulation is shown. The colors indicate the pseudopotential level as a function of the position. Between the electrodes with a height of about 1.2 eV, are the saddle points. The given simulation uses a 30 MHz trap drive with an amplitude of 100 V applied between the ring and the end-cap electrodes. If an ion obtains more energy than the saddle points, it can escape the trap. The trap depth is then the minimum energy the ion needs before it could either escape the trap, or hit one of the electrodes. For the trap depicted here, the required energy to impact the electrodes is above the saddle point energy and so the trap depth ϕ_D is simply the saddle point energy. The trap-depth efficiency factor κ_d can then be calculated with equation 2.36. For this geometry, the trap-depth efficiency factor κ_d is about 27%.

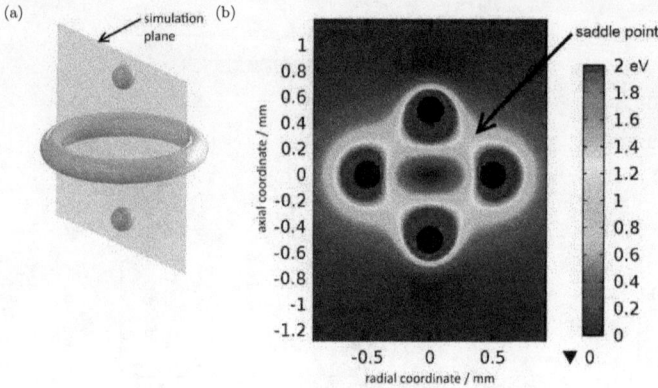

Figure 2.4: The geometry of a ring trap (a) and the simulation of the pseudopotential field in the simulation plane (b). The saddle points in the pseudopotential field give the trapping depth the ion experiences. The electrodes are about 1 mm apart on-center and 200 μm in diameter. An RF potential of amplitude 100 V and 30 MHz is applied between the ring and electrode. The resulting pseudopotential is plotted in eV.

2.2 Electronic States

Quantum information experiments can use electronic states of an ion to store a qubit. Many ion trap experiments use an alkaline earth element (eg. Ca, Sr, Ba) and for the experiments described here, the isotope ^{40}Ca is used. The electron configuration of neutral calcium (Ca I) is [Ar] $4s^2$, so that after single ionization (Ca II), the last unfilled electron shell $4s$, has just one valence electron. Ca II, with a single electron in the outer shell, then provides an alkali-like atomic structure which can be manipulated with optical radiation.

This section is organized as follows. First the spectroscopic notation used to describe the electronic states is reviewed (2.2.1). The Zeeman substructure of ^{40}Ca$^+$ is then described (2.2.2). Finally, the way the electronic states can be mapped to a qubit is then given (2.2.3).

2.2.1 Spectroscopic Notation

Describing the electronic states of atomic ions requires an understanding of spectroscopic notation, which is reviewed here briefly 2.2.1. A more detailed description can be found in references on atomic physics [61, 62].

The possible states of an alkali atom are described by several quantum numbers. The first is the principal quantum number n, describing the radial energy of the state. The next quantum number is its orbital angular momentum L as follows: for S-orbitals $L = 0$, for P-orbitals $L = 1$, and for D-orbitals $L = 2$. Representing the total spin of the electrons is the quantum number S, which for an atom such as Ca II, with one unpaired electron, is $1/2$. Ca II has a doublet splitting [63] due to it having a single valence electron. The doublet splitting is caused by whether the spin S is aligned or anti-aligned with the direction of the orbital angular momentum L. Since the spin of ^{40}Ca$^+$ is $S = 1/2$, this gives rise to a total angular momentum J, where J can take on the values $L + S$ or $|L - S|$. Except for the ground state S ($L = 0$), J can then have two values for each L, which gives rise to the doublet fine-structure splitting. The electronic state of an alkali ion or atom is then written as

$$n^{2S+1}\boldsymbol{L}_J, \tag{2.37}$$

where n is the principal quantum number, $\boldsymbol{L}$ is the electron's angular orbital name (S, P, D, F, etc), the multiplet splitting superscript $2S + 1 = 2$ indicates, that for non-zero angular momentum orbitals, the fine structure has doublets, and J is the total angular momentum. For instance $3\,^2$D$_{3/2}$ corresponds to an electronic state with a principal quantum number n of 3, an orbital angular momentum number L of 2 with the spin anti-aligned to the orbital angular momentum so that the total angular momentum number J is 3/2. The level structure for ^{40}Ca$^+$ is shown in figure 2.5. It shows the lowest empty electronic states and the visible or near-visible wavelength radiation, which is needed to perform the transitions between these states. The principal quantum numbers of the levels and multiplet splitting superscripts have been omitted for brevity. Coherent sources for such wavelengths can be obtained in the form of commercial diode-based lasers. Details of such sources are found in section 4.2.

An illustration of how to interpret this figure is as follows: Suppose that the ion starts in the electronic ground state S$_{1/2}$. Illuminating it with coherent light of wavelength near 396.8 nm drives the ion into the P$_{1/2}$ electronic state. In the absence of external radiation, this state will spontaneously decay, by emitting a photon with a characteristic lifetime of 7.1 ns. The P$_{1/2}$ state can decay into one of two different states; either the ground state S$_{1/2}$ (emitting a 397 nm photon), or the D$_{3/2}$ *metastable* state (emitting a 866 nm photon). The term metastable is used because a decay time constant of about a second is considered a long time in atomic physics, where most life times are measured in nanoseconds.

2.2.2   Zeeman Substructure

Each one of the electronic states of ^{40}Ca (S$_{1/2}$, P$_{1/2}$, P$_{3/2}$, D$_{3/2}$, D$_{5/2}$) has Zeeman substructure. For even numbered isotopes such as ^{40}Ca, the presence of a magnetic field removes the degeneracy of these energy levels. This is due to the coupling of the electron's total angular momentum J with the external static magnetic field, B_0. The levels of the Zeeman manifold are designated by the magnetic quantum number m_F, which corresponds to the projection of the total angular momentum in the direction of the magnetic field. It can take on values from $-J$ to $+J$, at unity intervals. With a stable static magnetic field in a particular direction, which is termed the *quantization axis*, the Zeeman splitting $\delta\omega_Z$ in even numbered isotopes is given by

$$\delta\omega_Z = m_F g_j \mu_B B_0 / \hbar, \tag{2.38}$$

where μ_B is the Bohr magneton, B_0 is the applied magnetic field, and the Landé factor g_j is dependent upon the electronic state. Table 2.1 contains these details of the Zeeman sublevels for each electronic state, and gives the computed splitting from an applied magnetic field.

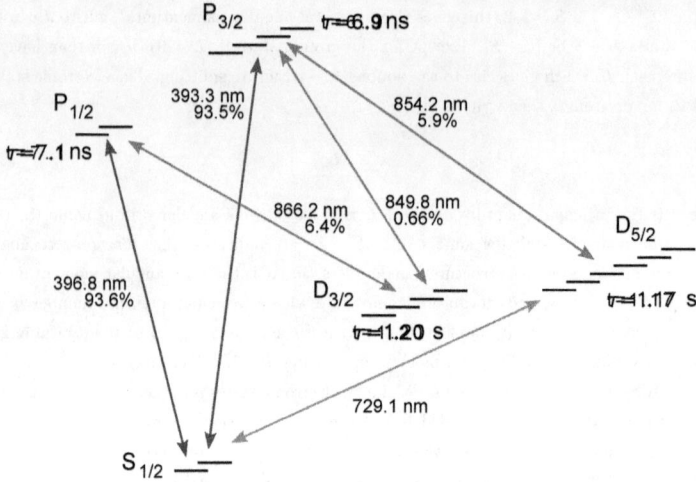

Figure 2.5: Relevant electronic structure of ^{40}Ca$^+$ ions. The qubit is typically mapped to the $S_{1/2}$ and $D_{5/2}$ states. Wavelengths of the transitions, lifetimes and branching ratios of the states have been taken from several sources [63–67].

state	g_j	splitting $\delta\omega_Z/$ $(2\pi\times\text{MHz/G})$
$S_{1/2}$	2	2.82
$P_{1/2}$	2/3	0.94
$P_{3/2}$	4/3	1.88
$D_{3/2}$	4/5	1.13
$D_{5/2}$	6/5	1.69

Table 2.1: The Zeeman substructure parameters for ^{40}Ca$^+$ electronic levels. The electronic state is given on the left, followed by the Landé factor g_j, and the computed splitting is given in terms of the applied magnetic field in Gauß (100µT).

The form of equation 2.38 helps to reveal why high-fidelity experiments with ^{40}Ca$^+$ require a magnetic field B_0. If the Zeeman splitting due to the magnetic field B_0 is not much larger than the frequency of any magnetic field noise present in the system, the direction of the quantization axis wanders non-adiabatically. This would then cause mixing between these states, leading to a loss of control over the Zeeman state. Although, transitions between the electronic states have not yet been described, they depend critically upon the Zeeman sublevel m_F. So that to faithfully control the electronic states, a magnetic field must be applied.

2.2.3 Electronic Qubit

The choice of the states, which encode the qubit, is called the qubit *basis* and often two states which have a long natural lifetime are used. For instance, the ground state ($S_{1/2}$) and the low lying metastable state ($D_{5/2}$) can be used, with a mapping of the D state to $|0\rangle$ and the S state to $|1\rangle$. Occasionally, the qubit is written with the electronic state as the label in Bra-Ket notation as $|S\rangle$ or $|D\rangle$. Since the ground state S is stable and the D state has a lifetime of about one second, the qubit then has a lifetime of about 1 second.

2.3 Quantized Motion of the Ion

The motion of the trapped ion also allows for quantum states to be stored. But most importantly, because of their large dipole moment, these motional states can be used to transfer information between the ions. This section first approximates the ion's stationary states using a quantum harmonic oscillator (QHO) (2.3.1) and reviews the solution in the position basis (2.3.2) and the energy basis (2.3.3). The ladder operators (2.3.4) used to relate the stationary states to each other are then introduced. The way two ions can move in a harmonic potential, their motional modes, is then described (2.3.5). Finally, the time dependent nature of the trapping fields is taken into account and the result is compared to the QHO simplification (2.3.6).

2.3.1 Quantum Harmonic Oscillator (QHO)

Using the pseudopotential approximation (see sec. 2.1.3), the motion of an ion in a quadrupole ion trap (QIT) can be treated as a QHO. The quantum harmonic oscillator is reviewed here for reference. The canonical quantization of a harmonic oscillator proceeds by treating the momentum and position variables in the classical Hamiltonian as quantum operators. An ion in a QIT can be described by three decoupled equation of motion. Considering just the x-direction and using the pseudopotential approximation, an ion in a harmonic trap with mass m and a trap frequency ω_x, the total energy or Hamiltonian is given by

$$\hat{H} = \frac{\hat{p}^2}{2m} + \frac{1}{2}m\omega_x^2\hat{x}^2, \tag{2.39}$$

where using the pseudopotential approximation for a QIT, the trap frequency is given by

$$\omega_x = \frac{\Omega_{rf}}{2}\sqrt{a_x + q_x^2/2}. \tag{2.40}$$

2.3.2 QHO Solution in Position Coordinates

The eigenfunctions in the position basis for a QHO are,

$$\psi_n(x) = \frac{\left(\frac{m\omega_x}{\pi\hbar}\right)^{1/4}}{\sqrt{2^n n!}} \exp\left(-\frac{mx^2\omega_x}{2\hbar}\right) H_n\left(x\sqrt{\frac{m\omega_x}{\hbar}}\right) \tag{2.41}$$

where $\psi_n(x)$ is the probability amplitude wave function at position x, $\hbar$ is the reduced Planck's constant, n is the motional quantum number (also called the number of *phonons*) of the particular eigenstate of the QHO and H_n are the Hermite polynomials. The root mean squared (RMS)

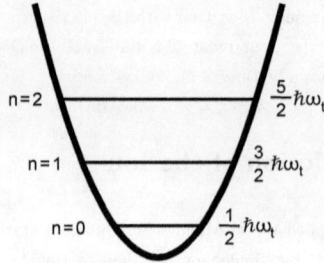

Figure 2.6: The Fock states (left) and the corresponding energy levels of the quantum harmonic oscillator. The energy levels are equally spaced and quantized. The lowest energy level has a non-zero amount of energy.

extent of these eigenstates a_n is given by

$$a_n = (n + 1/2)^{1/2} \sqrt{\frac{\hbar}{m_l \omega_x}}. \tag{2.42}$$

For instance, the RMS extent of the ground state wave function $a_0 = \sqrt{\hbar/(2m_l\omega_x)}$.

In general any state of the QHO can be written as a linear combination of these eigenstates. Figure 2.6 shows the energy structure of the lowest three ($n = 0, 1, 2$) eigenstates. States described by just one of these eigenfunctions, called *Fock* states, have a phonon number given by n, and an energy $E_n = \hbar\omega(n + 1/2)$. In the Schrödinger picture, only the phase of the state changes in time, so that the time dependence of these eigenstates is

$$\psi_n(x, t) = \exp\left[x, -i\hbar\omega_x t \left(n + \frac{1}{2}\right)\right] \psi_n(x). \tag{2.43}$$

These states form a complete basis so that any state of the QHO can be described in terms of a superposition of these eigenstates. A time dependent state, $\Psi(x, t)$, can be written as,

$$\Psi(x, t) = \sum c_n(t)\psi_n(t), \tag{2.44}$$

where $c_n(t)$ are the time changing amplitudes of the various eigenstates for the general state $\Psi(t)$. In the absence of coupling the QHO to another system, the c_n would be constant in time.

2.3.3   Dirac Bra/Ket Notation

Instead of working with probability amplitude functions in the position basis, the quantum states can be written using bra-ket notation. Any states of the QHO can then be represented by a ket, $|\Psi\rangle$, which can then be written as a sum of eigen-kets,

$$|\Psi\rangle = \sum_n c_n |n\rangle, \tag{2.45}$$

where $|n\rangle$ represents the n-th energy eigenstate of the QHO. The Hamiltonian can then be written as,

$$\hat{H} = \sum_n c_n \left(\hat{n} + \frac{1}{2} \right) \hbar \omega_{\mathrm{x}}, \tag{2.46}$$

where $\hat{n}$ is the number operator which measures the number of photons in the state.

2.3.4 Ladder Operators

To relate the different Fock states, the ladder operators $\hat{a}$ and $\hat{a}^\dagger$ can ease computations involving QHOs. The ladder operators related the different eigenstates as follows,

$$|n+1\rangle = \frac{\hat{a}^\dagger}{\sqrt{n+1}} |n\rangle \tag{2.47}$$

$$|n-1\rangle = \frac{\hat{a}}{\sqrt{n}} |n\rangle \tag{2.48}$$

$$\hat{a} |0\rangle = 0 \tag{2.49}$$

$$|n\rangle = \frac{(\hat{a}^\dagger)^n}{\sqrt{(n+1)!}} |0\rangle \tag{2.50}$$

$$\langle n| = \langle 0| \frac{\hat{a}^n}{\sqrt{(n+1)!}}. \tag{2.51}$$

For the ladder operators the commutation relation $[\hat{a}, \hat{a}^\dagger] = 1$ also holds. The operators for position $\hat{x}$ and momentum $\hat{p}$ can be written in terms of the ladder operators as

$$\hat{x} = \sqrt{\frac{\hbar}{2m\omega_{\mathrm{x}}}} (\hat{a}^\dagger + \hat{a}) \tag{2.52}$$

$$\hat{p} = i\sqrt{\frac{m\omega_{\mathrm{x}}\hbar}{2}} (\hat{a}^\dagger - \hat{a}). \tag{2.53}$$

The number operator $\hat{n}$ can be written in terms of the ladder operators as

$$\hat{n} = \hat{a}^\dagger \hat{a}. \tag{2.54}$$

2.3.5 Motion of Two Ions

If multiple ions are trapped in a three dimensional (3D) QHO, the motion of the ions is complicated by the additional degrees of freedom contributed by each ion. Consider a 3D QHO, where the principal axes x, y and z are non-degenerate, so that they each have distinct trapping frequencies ω_x, ω_y and ω_z. For a quadrupole ion trap with a cylindrical trapping potential, in the absence of static trapping fields, ω_z is twice the frequency of the two radial frequencies ω_x and ω_y (see section 2.1.2). If degeneracy of the radial trapping frequencies is removed so that the trapping frequency ω_x is lower than ω_y, a minimal energy configuration of two ions is one where they align themselves along a line in the x direction.

2.3.5.1 Modes of Motion for Two Ions

The motion of the two ions is coupled via the Coulomb force between them. One can then consider the motional *modes* of the Coulomb force coupled system. For each of the modes, an uncoupled QHO equation and associated annihilation and creation operators can be written.

The motional modes for two ions aligned along the x-axis are depicted in figure 2.7. Along the x direction, the motional modes are called "stretch" and x-"center of mass" (COM) modes. Perpendicular to the x-axis the modes are called are the y or z-axis "rocking" and COM modes. For multiple ions trapped in a single harmonic trapping potential, the relationship between the motional frequencies of these modes is fixed [40, 68]. For instance, the stretch-mode frequency ω_s is $\sqrt{3}$ times the COM-mode motional trap frequency ω_x.

The presence of an overall anharmonic potential causes these frequency relationships to no longer be valid [69, 70]. For instance, the addition of a quartic term to the harmonic potential can be used to create a double-well potential. Such a double-well model is analyzed below in section 3.4. The 2D arrays of trapped ions discussed in this thesis use a globally anharmonic potential to create an array of locally harmonic wells for each trapping site. If only a single ion is trapped in each well of the 2D array, one can consider this as a crystal of ions in an anharmonic trapping potential. In a simple configuration of just two neighboring wells, the spatial relationship between the ion's motion for each named mode in figure 2.7 is the same, while the frequency relationship will, in general, vary depending on how the trap is operated. For the rest of this thesis, the nomenclature for the relative motion of the ions shown in figure 2.7 (*stretch*, *rocking* and *COM*) will also be used for two neighboring potential wells in a 2D array, each containing a single ion.

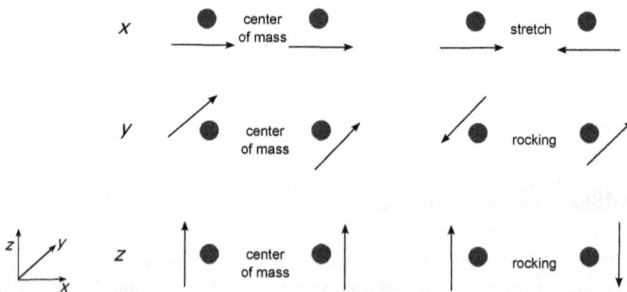

Figure 2.7: The motional modes of two ions in an ion trap. There are 6 motional modes to consider.

2.3.5.2 Phonon Bus

Because the quantized motion of the ions can be written in the coupled basis of these shared motional modes, each of the motional modes is a kind of communication channel called a *phonon bus*. This allows one to share a quantum state between the ions. The phonon bus and laser spectroscopy techniques (described below in section 2.4) have been used to create entangled states with up to 14 ions [71], each of which encodes a qubit.

2.3.6 Quantized Motion in a Quadrupole Ion Trap

While for many applications, using the pseudopotential approximation and treating the trapped ion as a QHO is sufficient, the time dynamics of the trapping fields must be considered for many experiments. Until this point, the Schödinger picture has been used. In this picture the eigenstates accumulate a phase with time, and the position and momentum operators are stationary in time. In following sections, the Heisenberg picture is used, where the eigenstates are constant in time, but the momentum and position operators accumulate phase dependent upon the energy. This allows one to more easily see the effect of the time varying fields on the eigenstates of the system. The motional hamiltonian $\hat{H}^{(m)}$ for an ion in a potential field U is given by

$$\hat{H}^{(m)} = \frac{\hat{p}^2}{2m_{\mathrm{I}}} + U(t, \hat{x}), \tag{2.55}$$

where $\hat{p}$ is the momentum operator, $\hat{x}$ is the position operator, m_{I} is the ion's mass, and $V(t, \hat{x})$ is the time varying and position dependent potential.

The Hamiltonian of an ion in a quadrupole ion trap (QIT) can be quantized with nearly the same method as a QHO [72, 73]. However, the states of an ion in a QIT which are analogous to the eigenstates of the QHO are called *quasistationary states*. Due to the oscillating trap drive field, these states are constantly exchanging energy with the external driving field, so that the wave function is breathing with the frequency of the trap drive Ω_{rf}. Considering just the x-direction, the Hamiltonian $\hat{H}^{(m)}$ of the motion of an ion in a trap is given by Leibfried et al. [73–75] as,

$$\hat{H}^{(m)} = \frac{\hat{p}^2}{2m_{\mathrm{I}}} + \frac{m_{\mathrm{I}}}{2} W(t)\hat{x}^2, \tag{2.56}$$

The time varying quantiy $W(t)$ is proportional to the oscillating trapping potential equal to

$$W(t) = \frac{\Omega_{\mathrm{rf}}^2}{4}[a_x - 2q_x \cos(\Omega_{\mathrm{rf}}t)], \tag{2.57}$$

where a_x and q_x are the trapping parameters described in section 2.1.2 and Ω_{rf} is the trap drive frequency. Using the solution from the Mathieu equations one can find a solution with a correspondence to the quantum harmonic oscillator. In the Heisenberg picture, the position and momentum operators of a particle in this time dependent trapping field are

$$\hat{x}(t) = \sqrt{\frac{\hbar}{2m_{\mathrm{I}}\nu}} \left[\hat{a}u^*(t) + \hat{a}^\dagger u(t) \right] \tag{2.58}$$

$$\hat{p}(t) = \sqrt{\frac{\hbar m_{\mathrm{I}}}{2\nu}} \left[\hat{a}\dot{u}^*(t) + \hat{a}^\dagger \dot{u}(t) \right],$$

where ν is called the reference oscillator and is equal to the secular frequency in the lowest order approximation, given in equation 2.19. The annihilation and creation operators here work on so-called quasistationary states, which are the dynamic counterpart to a QHO's Fock states. In the Schrödinger picture, these states have the form

$$|n, t\rangle = \exp\left[-i(n + 1/2)\nu t\right]|n\rangle (t), \tag{2.59}$$

where the time dependent phase progression $\exp(-i(n + 1/2)\nu t)$ multiplies the quasistationary states $|n\rangle (t)$. But we note here that, while in the QHO the stationary state's phase progresses with the trap's motional frequency, in a QIT, the quasistationary state's phase progresses with the reference oscillator's frequency ν. For processes that span a time, which is many multiples of the trap's period $2\pi/\omega_x$, any small difference between ν and the secular frequency could be noticed. With this in mind, from now on, the reference oscillator's frequency ν is approximated by the relevant secular motional frequency $\omega_x, \omega_y,$ or ω_z. The quasistationary states $|n\rangle (t)$, though they are not constant in time, have the property that they repeat themselves with the frequency of the trap drive Ω_{rf},

$$|n\rangle (t) = |n\rangle (t + 2\pi/\Omega_{rf}). \tag{2.60}$$

At intervals equals to the time period of the trap's RF drive, the trapped ion's quantum dynamics are then like the familiar QHO. In complete analogy to a standard QHO, the quasistationary states can be made with annihilation and creation operators [73]. As an example, the lowest order approximation of the probability amplitude wavefunction of the ground state of the ion in a QIT in the position basis is given by Leibfried et al. [73] as,

$$X_0(t, x) = \left(\frac{m\omega_x}{\pi\hbar}\right)^{1/4} \sqrt{\frac{1 - q_x/2}{1 + (q_x/2\cos(\Omega_{rf}t)}} \\ \times \exp\left(\left[i\frac{m\Omega_{rf}\sin(\Omega_{rf}t)}{2\hbar[2/q_x + \cos(\Omega_{rf}t)]} - \frac{m\omega_x}{2\hbar}\right]x^2 - i\frac{\omega_x t}{2}\right). \tag{2.61}$$

At moments in time equal to a multiple of the trap drive period $2\pi/\Omega_{rf}$, this expression simplifies to that of the canonical QHO's ground state wave function (see equation 2.41). For simplicity, we use the QHO description when possible, subject to the limitation that the dynamics are really only similar in the lowest order approximation and at times a multiple of the period of the trap drive.

For most of the rest of this thesis, the time dependent nature of the quasistationary states $|n\rangle (t)$ is not considered. Instead the simplification of a trapped ion as a QHO is used. All references to number or Fock states refer from now on to the quasistationary states. And we assume that we can substitute the reference oscillator frequency ν with the secular frequency and that the oscillations of the quasistationary states at frequency Ω_{rf} can be neglected. In the pseudopotential approximation, the QHO is the correct description of the motion of ions in a QIT, although it does not take into account the time dependent nature of the trapping fields. One notable exception to this simplification is for the interaction of the trapped ion with external fields. In this case, it can be necessary to take into account the time dependent nature of the trapping fields as is done below in section 2.4.

2.4   Laser Spectroscopy of Trapped Ions

As has been discussed in sections 2.2, the various motional states and long-lived electronic states can be used to store quantum information and a pair of states can store a qubit. In the absence of external electromagnetic (EM) fields, these states are well isolated from each other and form a type of quantum memory. But with the application of resonant radiation, typically in the form of a focussed laser beam, a coupling can be induced between the ion's electronic states or between motional states and electronic states of the trapped ion.

This section outlines the mathematical formalism used to describe such ion-light interactions. First the mechanisms of atom-light interactions are given (2.4.1). Then sideband spectroscopy of trapped atomic ions is described (2.4.2). Finally the method of electronic state preparation is given (2.4.3).

2.4.1   Atom Light Interactions

When coherent radiation is applied to an atom, which is resonant with an atomic transition, the atom's quantum state oscillates between the two states connected by the transition. The frequency at which two electronic states are coupled by the external coherent radiation is called the *Rabi* frequency Ω.

The terminology used to describe ion-light interactions with trapped atomic ions also includes the following terms. The frequency of the radiation to induce a coupling between electronic states without a change in the motional state is called the *carrier*. And the frequency of the radiation to induce an interaction between electronic states with a change in the motional states is called a *sideband*.

2.4.1.1   Two-level Approximation

To treat the interaction of applied electromagnetic fields and atoms mathematically, several simplifying assumptions are made, to which most experimental conditions are then restricted. First, a two level approximation is made. This means that only two levels of the ion's electronic structure are considered in the interaction with the light field. This is justified if the electromagnetic field used to drive the transition is resonant with, or close to, the carrier frequency ω_C and the rate of transitions to other off-resonant electronic levels is much smaller than the detuning δ of off-resonant transitions [73].

Generally, for optical transitions in ions these conditions are easily met, as the separation between optical transitions is more than 10^{12} Hz. However, when considering the Zeeman substructure, these conditions may not be so easy to meet (see sec. 2.6.2). For now however, consider just two of the ion's electronic energy levels, labelled $|e\rangle$, $|g\rangle$. The ion can then be described by a Hamiltonian $\hat{H}^{(e)}$ of the form,

$$\hat{H}^{(e)} = \hbar \frac{\omega_C}{2} \hat{\sigma}_z, \tag{2.62}$$

where the operator $\hat{\sigma}_z$ is the z-Pauli matrix. The Pauli matrices can be written in terms of the ground and excited states, in Dirac notation, as follows,

$$\hat{\sigma}_z = |e\rangle\langle e| - |g\rangle\langle g| \qquad\qquad \hat{\sigma}_x = |g\rangle\langle e| + |e\rangle\langle g| \qquad (2.63)$$
$$\hat{\sigma}_y = i(|g\rangle\langle e| - |e\rangle\langle g|) \qquad\qquad \hat{I} = |e\rangle\langle e| + |g\rangle\langle g|$$
$$\hat{\sigma}_+ = (\hat{\sigma}_x + i\hat{\sigma}_y)/2 \qquad\qquad \hat{\sigma}_- = (\hat{\sigma}_x - i\hat{\sigma}_y)/2.$$

2.4.1.2   Semi-classical Approximation

The second approximation made is that the interaction is treated semi-classically, so that the ion is quantized but the driving light field is not. The applied light field can either be absorbed by the ion, or stimulate emission. If the radiation is directed along the ion trap's x-axis, the coupling Hamiltonian $\hat{H}^{(i)}$ of the applied light field is given by,

$$\hat{H}^{(i)} = \hbar\frac{\Omega}{2}\hat{\sigma}_x \left[e^{i(k\hat{x}_S - \omega t + \phi)} + e^{-i(k\hat{x}_S - \omega t + \phi)} \right] \qquad (2.64)$$

where ω is the frequency of the applied light field, Ω is the Rabi frequency, $\hat{x}_S$ is the position operator of the ion, k is the wave number, and ϕ the phase of the applied field. The position of the ion $\hat{x}_S$ can affect the interaction by modulating the light field. The details of the laser frequency intensity and detuning and polarization and how this interacts with the ion's electronic state are all factors in the Rabi frequency Ω. For a dipole transition, the Rabi frequency Ω is given by,

$$\Omega = \frac{\langle g|(\boldsymbol{E_0} \cdot \hat{\boldsymbol{x}})|e\rangle\, 2q}{\hbar}, \qquad (2.65)$$

where $\boldsymbol{E_0}$ is the electric field amplitude of the driving light field and $\hat{\boldsymbol{x}}$ is the position vector operator of the ion.

To obtain the time dynamics of this applied field, one can use the interaction picture [76]. Consider the trapped ion's free Hamiltonian

$$\hat{H}_0 = \hat{H}^{(e)} + \hat{H}^{(m)}, \qquad (2.66)$$

where $\hat{H}^{(m)}$ is the motional Hamiltonian of the ion (eq. 2.56). The interaction $\hat{V} = \hat{H}^{(i)}$ is then applied. The free Hamiltonian is considered in the Heisenberg picture, so that the time evolution operator of $\hat{H}_0$ is $\hat{U}_0 = \exp[-(i/\hbar)\hat{H}_0 t]$.

2.4.1.3   Rotating-Wave Approximation

Next the rotating-wave approximation is applied. Only considering excitations of the lowest-order secular sidebands, and just considering the lowest-order approximation to the ion's motion, the interaction Hamiltonian $\hat{H}_{\text{int}} = \hat{U}_0^\dagger \hat{V} \hat{U}_0$ becomes

$$\hat{H}_{\text{int}}(t) = \frac{\hbar\Omega_0}{2} \left\{ \sigma_+ \exp\left[i\eta(\hat{a}e^{-i\omega_x t} + \hat{a}^\dagger e^{i\omega_x t})\right] e^{i(\phi - \delta t)} + \text{h.c.} \right\} \qquad (2.67)$$

where $\Omega_0 = \Omega/(1 + q_x/2)$ is the *effective Rabi* frequency [73], h.c. is the hermitian conjugate of the expression inside the curly brackets, and the *Lamb-Dicke* parameter η is given by

$$\eta = ka_0 \qquad (2.68)$$

where a_0 is the size of the ground-state wave function. The detuning $\delta = \omega - \omega_C$ is the difference between the applied laser frequency and the carrier frequency. The model given here only takes into account a one-dimensional model of the ion's motion. As such the form for the effective Rabi frequency Ω_0 does not include motion from other directions and modes. The modes of motion in these other directions are called *spectator* modes, the presence of which, can alter the effective Rabi frequency [77].

2.4.2   Sideband Spectroscopy

For detunings δ near to a multiple of the trap frequency ω_x, the applied laser field will couple the electronic states to the motional states. For a detuning $\delta = -\omega_x$, this is called the first *red sideband* (RSB), and when $\delta = \omega_x$ this is called the first *blue sideband* (BSB).

2.4.2.1   Lamb-Dicke Regime

If the extent of the ion's motion is much smaller than the wavelength of radiation so that $\eta^2(2\bar{n} + 1) \ll 1$, where $\bar{n}$ is the average motional quanta, then it is said to be in the *Lamb-Dicke regime*. In this case, the exponent in $\hat{H}_{\text{int}}$ can be expanded to lowest order in η to give

$$\hat{H}_{\text{int}}(t) = \frac{\hbar\Omega_0}{2} \left\{ \sigma_+ \left[1 + i\eta(\hat{a}e^{-i\omega_x t} + \hat{a}^\dagger e^{i\omega_x t}) \right] e^{i(\phi - \delta t)} + \text{h.c.} \right\} \qquad (2.69)$$

2.4.2.2   Red, Blue and Carrier Transitions

For the three resonances (RSB, BSB, and carrier transitions), the evolution of the two-level system can be calculated. For instance, for the detunings $\delta = (-\omega_x, 0, \omega_x)$, then in the two-level approximation, the interaction Hamiltonians $\hat{H}_{\text{int}}(t)$ becomes,

$$\hat{H}_{\text{rsb}}(t) \simeq \eta \frac{\hbar\Omega_0}{2} (\hat{a}\sigma_+ e^{i\phi} + \hat{a}^\dagger \sigma_- e^{-i\phi}), \qquad \delta = -\omega_x \qquad \text{RSB} \qquad (2.70)$$

$$\hat{H}_{\text{car}}(t) \simeq \frac{\hbar\Omega_0}{2} (\sigma_+ e^{i\phi} + \sigma_- e^{-i\phi}), \qquad \delta = 0 \qquad \text{carrier} \qquad (2.71)$$

$$\hat{H}_{\text{bsb}}(t) \simeq \eta \frac{\hbar\Omega_0}{2} (\hat{a}^\dagger \sigma_+ e^{i\phi} + \hat{a}\sigma_- e^{-i\phi}), \qquad \delta = \omega_x \qquad \text{BSB.} \qquad (2.72)$$

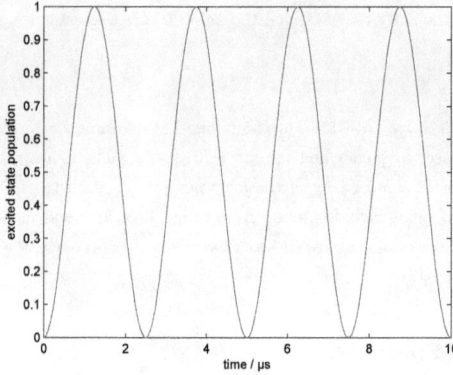

Figure 2.8: Probability of a two level ion being in excited state with applied electromagnetic radiation at the carrier frequency. Shown with an effective Rabi frequency, Ω_0, of $2\pi \times 4 \times 10^5$ rad/s.

2.4.2.3 Rabi Flops

The evolution of the ion's state $|\psi\rangle = c_g |g\rangle + c_e |e\rangle$ with induced carrier transitions, in the absence of spontaneous emission, is

$$\dot{c}_e = c_g e^{i\phi} \frac{\Omega_0}{2} \tag{2.73}$$
$$\dot{c}_g = -c_e e^{-i\phi} \frac{\Omega_0}{2}.$$

So that the state coefficients c_g and c_e, given the initial condition $c_g(0) = 1, c_e(0) = 0$, are

$$c_g = \cos(t\frac{\Omega_0}{2}) \tag{2.74}$$
$$c_e = \sin(t\frac{\Omega_0}{2}).$$

And the probability that the state will be in the ground state P_G or excited state P_E, is

$$P_G = |c_g(t)|^2 = \cos^2(t\frac{\Omega_0}{2}) = \frac{1 + \cos(\Omega_0)}{2} \tag{2.75}$$
$$P_E = |c_e(t)|^2 = \sin^2(t\frac{\Omega_0}{2}) = \frac{1 - \cos(\Omega_0)}{2}.$$

The solutions of these equations describe the well known phenomenon of Rabi flops [78]. The solution to these equations is shown in figure 2.8. One can see that the probability that the ion is in the excited state oscillates with an angular frequency of Ω_0. However, an important detail of this interaction is that while the probability of the ground or excited states is oscillating with the effective Rabi frequency Ω_0, the coefficients of the states c_g, c_e are oscillating with half the effective Rabi frequency. So that the ground state coefficient $c_g(t)$ at time t evolves to $c_g(t + T) = -c_g(t)$ at a time $t + T$, where $T = 2\pi/\Omega_0$.

2.4.3 Electronic State Preparation

For trapped ion quantum computation, the ion must be prepared in a well defined initial state. For the electronic state, the ground state is a natural choice. In order to prepare the ion in one of its Zeeman sublevels, laser light is applied in a process called optical pumping. Either a frequency selective method or a polarization selective method can be used as follows.

Circularly polarized laser light resonant with the $S_{1/2}$ to $P_{1/2}$ transition can repetitively pump the ion to one of its Zeeman sublevels. The laser light will excite the $S_{1/2}$, $m_F = 1/2$ to $P_{1/2}$, $m_F = -1/2$ transition. The ion can decay into either ground state, but the circularly polarized light cannot empty any populations which end up in the $S_{1/2}$, $m_F = -1/2$ state. In this way, the ion is optically pumped into the $S_{1/2}$, $m_F = -1/2$ state. Laser light at 866 nm repumps the ion should it decay into the $D_{3/2}$ state. This keeps the electronic state of the ion in a closed loop. If the polarization of the ion is not purely circularly polarized, then this process will not produce a perfectly prepared state.

A frequency selective process can also be used to prepare the initial electronic state [79]. Laser light (729 nm) resonant with the $S_{1/2}$, $m_F = 1/2$ to $D_{5/2}$, $m_F = -5/2$ quadrupole transition is first applied to an ion the ground state manifold. Laser light at 854 nm can then be used to excite the ion to the $P_{3/2}$, $m_F = -3/2$ state. Dipole selection rules dictate that this is the only allowable transition. The ion can then decay to either of the ground states. However, if the ion decays into the $S_{1/2}$, $m_F = -1/2$ state, this optical pumping stops, as the 729 nm laser light is not resonant with any transition from this state.

2.5 Photoionization of Calcium

In order to load the trap, ions need to be created within the trapping volume. A beam of neutral Ca I atoms are produced by a heated oven inside the vacuum chamber. To ionize these neutral atoms, a two-step photoionization process was used here [80–82]. The process, which uses two lasers is depicted in figure 2.9. The first laser at roughly 423 nm resonantly excites the $4s^2\,^1S_0$-$4s4p\,^1P_1$ transition, and a second light source, with a wavelength of roughly 390 nm or shorter excites it to a Rydberg state or directly ionizes it. The ion trap's electric fields can then ionize the Rydberg atom. Since the source of calcium inside the oven has a mixture of isotopes, an isotope selective ionization process is desirable. Because the first laser (423 nm) is a resonant transition, where the frequency is isotope dependent, this allows for just $^{40}Ca^+$ ions to be produced in the trap.

2.6 Laser Cooling

Trapped ions can be cooled with laser light [51, 83–85]. Laser cooling works because, when an atom absorbs or emits light, the light imparts a momentum to the atom. A vibrating trapped ion can then use the Doppler effect [86] of the motion of the ion relative to a laser beam to only

Continuum

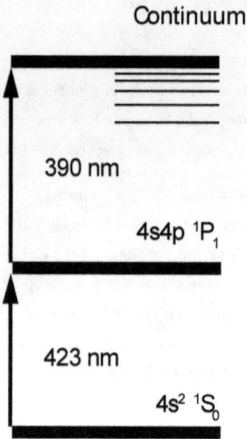

Figure 2.9: The relevant electronic levels for photoionization of neutral calcium (calcium I) to produce singly charged calcium (calcium II). The two-photon process depicted can either directly ionize calcium I or produce Rydberg states. The Rydberg atoms can then be ionized by the electric fields of the ion trap. A resonant laser at the 4s $^1S_0$-4s4p $^1P_1$ transition allows for the process to be isotope selective. A light source of roughly 390 nm can create Rydberg atoms or a laser with a wavelength shorter than 390 nm could directly ionize the atoms in the 4s4p $^1P_1$ state.

absorb the light when the ion is travelling towards the light and thereby have its speed reduced. This can give rise to an average damping force, which cools the ion.

There are two main regimes of laser cooling of trapped ions, which compare the natural linewidth of the laser cooling transition γ to the trap frequency ω_x. When the linewidth of the cooling transition is much larger than the trapped ion's motional frequencies, so that $\gamma \gg \omega_x$, is called the *weak binding* limit. Cooling in this regime is also called Doppler cooling. A quantitative treatment of Doppler cooling below takes into account the quantized electronic states of the ion, but treats the motion as a continuous variable.

Cooling in the resolved sideband regime [87], where the natural linewidth of the cooling transition is much less than the trapping frequency, so that $\gamma \ll \omega_x$, is also called the *strong binding* limit. Resolved sideband cooling allows a trapped ion to be cooled beyond the limits of Doppler cooling; to the quantum ground state of motion. A quantitative treatment of resolved sideband cooling below takes into account the quantized states of the trapped ion's motion.

A mathematical treatment, so as to understand the limits of Doppler laser cooling in the weak binding limit, follows (2.6.1). Next, the phenomena of dark states is described and a method to avoid them during Doppler cooling is given (2.6.2). Finally, resolved sideband cooling is then described (2.6.3).

2.6.1 Doppler Cooling

Doppler cooling works by illuminating the trapped ion with detuned laser light, so that's its frequency is less than the ion's electronic transition frequency (red shifted). The oscillations of the ion in the trap cause it to be travelling towards the laser source part of the time. When it is travelling towards the source, a Doppler shift [86] causes the light to increase in frequency from the ion's perspective, so that it can be resonant with the ion's transition. The absorption of the light then reduces the momentum in the direction the ion was travelling by an amount $\hbar k$. The light will be spontaneously re-emitted in a random direction. Since on average the random emission of a photon has no preferred direction, this process then cools the ion.

A theoretical approach following the methods of Stenholm [88] gives an understanding of laser cooling sufficient to build and operate an experiment. A number of simplifications are made with this approach, which are generally adhered-to during the operation of the experiment. First the laser light is treated semi-classically and the ion is treated as a two-level system, which we justified in section 2.4. The trap's time varying potential is also approximated by a pseudopotential which is assumed to be purely harmonic. This treatment fails to take into account important experimental phenomena such as dark-states, which we will treat separately in the next section.

Unlike the coherent sideband spectroscopy described in section 2.4.2. The excited states used for Doppler cooling usually have a short lifetime and therefore a correspondingly large linewidth, γ. And this linewidth is not much smaller than the trap frequency ω_x. The reason such a transition is used is because the cooling power is inversely proportional to the lifetime of the transition chosen. Also, the cooling process uses spontaneous emission to cool the ion and this process is inherently decoherent. After spontaneous emission, the state of the ion cannot necessarily be described by the pure state formalism that has been used up to this point. For these reasons, a semi-classical theory to describe Doppler cooling is introduced. First in this section, the density matrix formalism is introduced. Then the evolution of the ion's state under a light-ion interaction is given. This allows the excited state probability to be computed, so that the average force on the ion can be computed. This force has a damping behavior which slows the motion of the ion. Because of the spontaneous emission however, the ion receives kicks which heat the ion. The limit of Doppler cooling can then be calculated when the two effects of cooling and heating cancel.

2.6.1.1 Density Matrix Formalism

To describe the probabilistic mixed state of the ion due to spontaneous emission, the density matrix representation is introduced here. Considering just the electronic state of the ion. The density matrix representation of the state of an ion can written as a sum of pure states as follows,

$$\rho = \sum_n c_n |n\rangle \langle n|, \tag{2.76}$$

where $|n\rangle$ are the quantized electronic states of the ion. The coefficients c_n are probabilities of finding the ion in that state. This is an important conceptual difference from the quantum principle of superposition. Here, the quantum state is said to be in a mixture (as opposed to "pure"), so that the likelihood of finding the system in a particular pure quantum state $|n\rangle$ is simply proportional to the coefficient c_n. In contrast, if a quantum state $|\psi\rangle$ is in a superposition of states, so that $|\psi\rangle = \sum_n c_n |n\rangle$, then the likelihood of measuring the system in state $|n\rangle$ is proportional to $|c_n|^2$ and it is possible to measure interference between the states since the states are present at the same time. A density matrix is used to represent the probabilistic mixture of quantum states, averaged over many experimental cycles. For any single experiment, only one quantum state $|n\rangle$ is present with probability c_n.

2.6.1.2 Louiville-von Neumann Equation

Using this description, the evolution the density matrix subject to a Hamiltonian H evolves according to the von Neumann equation,

$$i\hbar\frac{d\rho}{dt} = -\frac{i}{\hbar}[\hat{H}, \rho]. \tag{2.77}$$

When an atom absorbs a photon, it receives a momentum kick in the direction the photon was travelling, and this is included in the above Hamiltonian. However, if it spontaneously emits a photon some time later, this is in a random direction. In order to add this random effect to the dynamics, the Liouville $L^d\rho$ is added to the von Neumann equation (notation from Cirac et al. [89, 90]).

$$L^d\rho = \gamma(2\sigma^-\tilde{\rho}\sigma^+ - \sigma^+\sigma^-\rho - \rho\sigma^+\sigma^-)/2 \tag{2.78}$$

where γ is the spontaneous emission rate and $\tilde{\rho}$ accounts for the momentum transfer in the spontaneous emission of a photon and is given by

$$\tilde{\rho} = \frac{1}{2}\int_{-1}^{1} dz e^{ik\hat{x}z}\rho e^{-ik\hat{x}z}3(1 + z^2)/4. \tag{2.79}$$

The form of $\tilde{\rho}$ above assumes that the spontaneously emitted photons have a dipole radiation profile. By expanding L^d in terms of the Lamb-Dicke parameter η, and taking the approximation that the laser cooling rate (change in the motional state) is slow compared to both the ion's motional frequency and the internal atomic frequencies, a perturbative approach can be taken to give the excited or ground state populations.

2.6.1.3 Cooling and Heating

Here, cooling is only considered along the x-direction of motion and the motion of the ion is considered classically with an x-direction velocity v_x. When the ion is illuminated by laser light, the electronic state reaches a quasi-stationary excited-state population, which is then given by Loudon [91] as

$$\rho_{ee} = \frac{(\Omega_0/2)^2}{\delta_{\text{eff}}^2 + (\gamma/2)^2 + \Omega_0^2/2} \tag{2.80}$$

where Ω_0 is the effective Rabi frequency, γ is the rate of spontaneous emission of the excited state, $\rho_{ee} = \langle e|\rho|e\rangle$ is the excited state population, and δ_{eff} is the effective detuning of the light. The effective detuning of the laser light is dependent upon the ion's velocity, which we assume to change much slower than the time it takes for the excited state probability to reach its quasi-stationary state.

The excited state can also be written in terms of the saturation parameter $s = |\Omega_0|^2/\gamma^2$, where Ω_0 is the effective Rabi frequency. Knowing the excited state population allows one to calculate the average scattering rate, and hence the average force on the ion. For small velocities, the effective detuning, due to the Doppler shift, can be linearized in $\boldsymbol{v}$ so that $\delta_{\mathrm{eff}} = \Delta - \boldsymbol{k} \cdot \boldsymbol{v}$, where $\Delta = \omega - \omega_{\mathrm{C}}$ is the nominal detuning of the laser. The average force in the x-direction $\bar{F}_x$ is then [73]

$$\bar{F}_x = F_0(1 + \kappa_{\mathrm{dc}} v_x), \tag{2.81}$$

where F_0 is the average radiation pressure in the x-direction that displaces the ion from the center of the trap, v_x is the velocity of the ion in the x-direction, and κ_{dc} is the damping coefficient proportional to the speed of the ion, given by

$$\kappa_{\mathrm{dc}} = \frac{8k\Delta/\gamma^2}{1 + s + (2\Delta/\gamma)^2}. \tag{2.82}$$

On average, the ion sees the above damping force, which would lead to a rate of change of the ion's energy in time of $\dot{E}_c = F_0 \kappa_{\mathrm{dc}} \langle v_x^2 \rangle$, where $\langle v_x^2 \rangle$ is the expected value of the square of the ion's velocity. However, laser cooling of the ion is a process which is limited by the random impulses it receives during spontaneous emission. Near the limit of Doppler cooling, in the low velocity limit, the heating rate due to these random kicks is then on average given by Leibfried et al. [73] as

$$\dot{E}_h \simeq \frac{1}{2m}(\hbar k)^2 \gamma \rho_{ee}(v = 0)(7/5). \tag{2.83}$$

When the ion reaches a steady state, so that the heating rate and the cooling rate equal each other, the final energy of the ion is then

$$E_{\mathrm{min}} = \frac{7\hbar\gamma\sqrt{1+s}}{20}. \tag{2.84}$$

where the laser detuning $\Delta = (\gamma\sqrt{1+s})/2$.

2.6.2   Dark States

The two level approximation is not able to explain a well-known phenomena. With certain laser intensities, polarizations and detunings, an atom can cease to fluoresce in ways that are not predicted by a simple two-level approximation [92, 93]. Because a $^{40}\mathrm{Ca}^+$ ion has a Zeeman sub-structure, with a splitting not much greater than the Rabi frequency of the Doppler cooling laser, the two-level approximation is not always valid. There exist electronic states of the ion which are eigenstates of the interaction Hamiltonian of the combined laser-ion system [94, 95]. Some such eigenstates do not have a significant population in the excited state and show reduced or no fluorescence.

In order to quantitatively treat this phenomena for $^{40}\mathrm{Ca}^+$, all eight Zeeman sub-levels of the electronic states $\mathrm{S}_{1/2}$, $\mathrm{P}_{1/2}$ and $\mathrm{D}_{3/2}$, as well as the laser parameters of the 397 nm and

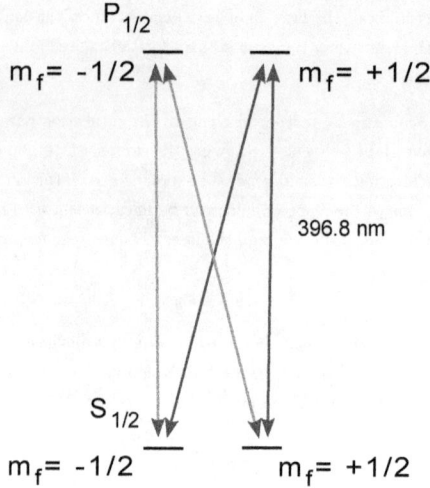

Figure 2.10: An illustration of dark states involving just the $S_{1/2}$ and $P_{1/2}$ Zeeman sublevels. Laser light can coherently excite the populations in each of the ground states to the two excited states. If the phase of the ground states is opposite, $|-\rangle = (|m_F = -1/2\rangle - |m_F = +1/2\rangle)/\sqrt{2}$, then they destructively interfere and the excited state population (and hence fluorescence) is reduced.

866 nm sources must be taken into account [94–96]. This includes the laser detunings, intensities and polarizations with respect to the magnetic quantization axis and its strength. This large parameter space is best treated numerically and is beyond the scope of this thesis. However, some insight into these phenomena can be gained by considering just four Zeeman sublevels of the $S_{1/2}$ and $P_{1/2}$ state, and considering this system as if it were two symmetric Lambda-level structures as described by Wynands et al. [97]. In figure 2.10 is a depiction of these four Zeeman sublevels in the absence of a magnetic field. Linearly polarized laser light at 397 nm will excite both Zeeman sublevels of the $S_{1/2}$ state to the one of the $P_{1/2}$ sub-levels. If the initial state of the ion is a coherent superposition of these ground states, a convenient basis is

$$|+\rangle = \frac{|m_F = -1/2\rangle + |m_F = +1/2\rangle}{\sqrt{2}} \tag{2.85}$$
$$|-\rangle = \frac{|m_F = -1/2\rangle - |m_F = +1/2\rangle}{\sqrt{2}}$$

If the ion is initially in the $|+\rangle$ state, then excitation will bring the ion into the $P_{1/2}$ levels. However, if the ion is in the $|-\rangle$ state, exciting both ground states at the same time will cause destructive interference because they have the opposite phase. In this case $|-\rangle$ is said to be a *dark state* because it exhibits reduced or no fluorescence compared to $|+\rangle$. Application of a magnetic field will causes these levels to split so that the states $|+\rangle$ and $|-\rangle$ will oscillate between each other as time progresses. In this way, an applied magnetic field can be used to destabilize the dark state.

The dark states of $^{40}\text{Ca}^+$-like ions have been studied and they can be avoided by creating a

Zeeman splitting with the magnetic field quantization axis at the correct angle to the polarization of the Doppler cooling beams [96]. The following guidelines are recommended: For ^{40}Ca$^+$ ions Doppler cooled without substantial power broadening, the angle of linearly polarized 397 nm and 866 nm light should be perpendicular with respect to the magnetic quantization axis. The 866 nm light should be blue detuned half the excited state's linewidth. And the strength of the magnetic field should be such that is has a splitting of between 0.01 and 0.1 of the excited state's linewidth [96].

2.6.3   Sideband Cooling

In the strong binding limit, where the linewidth of the cooling transition is much smaller than the trap's motional frequencies, the ion can be cooled to the motional ground state [98]. A semi-classical description is as follows, where the cooling laser is directed along the x axis. Consider the laser frequency ω_I in the trapped ion's oscillating reference frame. Because of the first order Doppler effect, the laser's frequency ω is frequency modulated at the motional frequency of the trapped ion ω_x. Specifically, it has wideband frequency modulation (FM) with a sinusoidal signal [99], so that the laser frequency of the laser in the ion's reference frame can be written as

$$\omega_I = \omega + \omega_x \eta (2\bar{n} + 1) \sin(\omega_x t) \tag{2.86}$$

where η is the Lamb-Dicke parameter and $\bar{n}$ is the average phonon number. The FM modulation index β_{FM} [99] is then equal to the quantity $\eta(2\bar{n} + 1)$. The spectrum of such a signal has a carrier at the laser's laboratory frequency ω, with multiple sidebands spaced at ω_x. The frequency modulated electric field $\boldsymbol{E}(t)$ of the cooling laser can then be written as

$$\boldsymbol{E}(t) = \boldsymbol{E}_0 \sum_{n=-\infty}^{\infty} J_l(\beta_{FM}) \cos\left[(\omega + l\omega_x)t\right] \tag{2.87}$$

where $\boldsymbol{E}_0$ is the amplitude of the laser's unmodulated electric field at the ion and $J_l(\beta_{FM})$ are Bessel functions of the first kind, of order l. The intensity of the l-th sideband of the laser is then proportional to $J_l^2(\beta_{FM})$. At high temperatures (large $\bar{n}$), there are many sidebands with sufficient intensity, so that practical resolved sideband cooling would be technically difficult. For $\beta_{FM} \ll 1$, the properties of these Bessel functions give that most of the laser intensity is contained in the carrier and the first order side bands. The intensity of the first order sidebands relative to the carrier is then equal to twice the square of the first order Bessel function of the first kind $2J_1^2(\beta_{FM})$, which, in the lowest order approximation, is equal to $\beta_{FM}^2/2$. Once the ion is cool enough (usually by the application of Doppler cooling) so that $\beta_{FM}^2 \ll 1$, in the lowest order approximation, the intensity of the carrier is approximately $1 - \beta_{FM}^2/2$. This is the same as the condition to be inside the Lamb-Dicke regime (see sec. 2.4.2.1) and it means that the first order Doppler shift can be effectively ignored.

Resolved sideband cooling can then proceed as follows using sideband spectroscopy (see sec. 2.4.2). An ion in the ground state is given a π pulse on the red sideband (RSB), so that the laser frequency $\omega = \omega_C - \omega_x$, where ω_C is the electronic transition carrier frequency. Because the ion is inside the Lamb-Dicke regime, the intensity of the laser light at the RSB is not appreciably

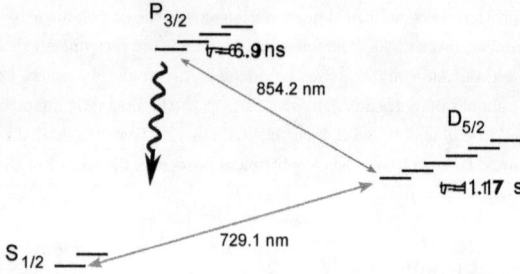

Figure 2.11: Sidband cooling of a ^{40}Ca$^+$ ion. After Doppler cooling and electronic state preparation to the $S_{1/2}, m_F = -1/2$ state, the application of a frequency selective RSB pulse can transfer the ion's state from $S_{1/2}, m_F = -1/2$ to $D_{5/2}, m_F = -5/2$. Application of the 854 nm laser can then transfer the D-state population to $P_{3/2}, m_F = -3/2$ since this is the only dipole-allowed transition. The excited $P_{3/2}, m_F = -3/2$ then can only decay to the $S_{1/2}, m_F = -1/2$ state. If the Lamb-Dicke parameter η is small, then the 854 nm transition to $P_{3/2}$ and the spontaneous decay to the ground state are both unlikely to increase the motional quantum number, and each cycle of this process should decrease the average phonon number by 1. Such a process can produce an ion in the motional ground state with more than 99.9% probability [100].

modulated to the carrier or other sidebands. This π pulse on the RSB raises the electronic state, but reduces the ion's motional state by one (see sec. 2.4.2). Spontaneous emission can then follow, bringing the ion again into the electronic ground state. Usually, spontaneous emission does not change the motional state (with probability $\sim (1 - \eta^2)$. This process can be repeated until the ion is in the ground state.

For ^{40}Ca$^+$, the 729 nm S to D transition, with a natural linewidth of about 1 Hz, offers an easily resolved sideband transition. However, to speed up the process, the ion is pumped out of the long-lived state D with 854 nm laser light. Figure 2.11 depicts one method as follows: After Doppler cooling and electronic state preparation to the $S_{1/2}, m_F = -1/2$ state, the application of a frequency selective RSB pulse can transfer the ion's state from $S_{1/2}, m_F = -1/2$ to $D_{5/2}, m_F = -5/2$. Application of the 854 nm laser can then transfer the D-state population to $P_{3/2}, m_F = -3/2$ since this is the only dipole-allowed transition. The excited $P_{3/2}, m_F = -3/2$ then can only decay to the $S_{1/2}, m_F = -1/2$ state. If the Lamb-Dicke parameter η is small, then the 854 nm transition to $P_{3/2}$ and the spontaneous decay to the ground state are both unlikely to increase the motional quantum number, and each cycle of this process should decrease the average phonon number by 1. Such a process can produce an ion in the motional ground state with more than 99.9% probability [100].

2.7   Qubit Rotations and Gates

Up to this point, we have discussed how to prepare and manipulate the quantum states of ions. In order to leverage these capabilities into a quantum simulator or computer, the simulation's interactions or the computational algorithm needs to be encoded into a set of operations which manipulate the states of ions. For simulations this is often done by preparing a state and then using pulses of laser light to simulate the desired system dynamics. For computation, the algorithm is usually compiled into a series of laser pulses which process the qubit(s). For instance if a qubit, $|\Psi\rangle$, is encoded into the ions's electronic states (for instance $|S\rangle = |1\rangle$ and $|D\rangle = |0\rangle$), then the state can be written as

$$|\Psi\rangle = \left[\cos\left(\frac{\theta_0}{2}\right)|0\rangle + e^{i\phi_0}\sin\left(\frac{\theta_0}{2}\right)|1\rangle\right], \qquad (2.88)$$

where θ_0 and ϕ_0 are both real numbers which describe the state of the qubit. The angles θ and ϕ can be represented as the angle of a unit vector from the vertical z-axis and the azimuthal angle in the xy-plane as shown in figure 2.12. Such a representation is called the *Bloch sphere* representation of the qubit. In the rest of this section, first single qubit gates are described (2.7.1), and then an overview of multi-qubit gates is given (2.7.2).

2.7.1 Single Qubit Rotations

Usually, the qubit is considered in the Heisenberg representation, so that in the absence of external fields, it does not rotate. For the S-D qubit above, a pulse of 729 nm laser light incident upon the ion, resonant with the $S_{1/2}$-$D_{5/2}$ transition will rotate $|\psi\rangle$ as [101, 102],

$$R(\theta, \phi) = \exp\left(\frac{i}{\hbar}\hat{H}_{car}\theta\right). \qquad (2.89)$$

Here $\hat{H}_{car}$ is the carrier Hamiltonian (eq. 2.71), θ and ϕ point in the direction of the xy-plane about which the rotation takes place. The magnitude of the rotation is given by the length of the laser pulse times the Rabi frequency (i.e. $\theta = \Omega_0 t$). These type of on-carrier rotations are called single qubit rotations and they are further subdivided into rotations about the x or y axis as follows

$$\begin{aligned} R_x(\theta) &= R(\theta, \pi) & R_{\bar{x}}(\theta) &= R(\theta, 0) \\ R_y(\theta) &= R(\theta, 3\pi/2) & R_{\bar{x}}(\theta) &= R(\theta, \pi/2) \end{aligned} \qquad (2.90)$$

where the subscripts $\bar{x}$ and $\bar{y}$ mean that the rotations are around the negative x and y axis respectively. While these rotations allow one to perform any action on a single qubit, for efficiency's sake, rotations around the z-axis can be performed by the application of a far off-resonant beam which produces an AC-stark shift [103].

2.7.2 Quantum Logic Gates

For coupling between the motional qubit and the electronic qubit, application of RSB and BSB pulses allows the excitation or depopulation of motional Fock states. Since the motional states of the ions are coupled to each other through the Coulomb force, the electronic state of one ion can then be transferred to motional states and used to manipulate the electronic states of other ions.

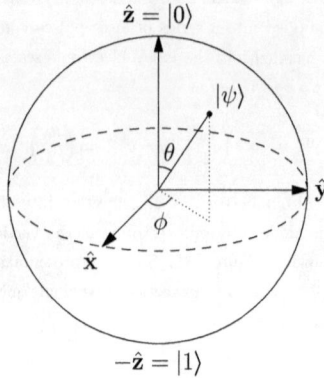

Figure 2.12: Bloch sphere representation of a qubit showing the relation between the angles θ and ϕ to a vector on the unit sphere, which represents the qubit $|\psi\rangle$.

It turns out that almost any two qubit interaction is enough to form a universal basis of quantum gates [27]. In most of the quantum information experiments with trapped ions, a linear quadrupole ion trap is used to store a string of ions. The motional mode of the string is shared between all the ions, so that it forms a phonon bus, which allows for a shared qubit between the ions in the string. Two qubit interactions such as the Cirac-Zoller CNOT gate [104], the Mølmer-Sørensen-type controlled-phase-gate [105] or even fault tolerant gates [106] have been used, along with single qubit rotations, to create a universal set of operations. For $^{40}Ca^+$ ions, quantum logic gates can be done with lasers in roughly a few 10s of µs [107].

An alternative proposal, which we aim to implement, uses ions in a 2D array of microscopic traps, with nearest neighbor interactions [108]. The interactions described above, so far, have not been implemented in 2D arrays of ions traps. A proposed method to create pairwise entanglement and perform two-qubit gates in 2D arrays is given in the next chapter in sections 3.2.6 and 3.2.8.

2.8 Electronic State Measurement

One of the areas in which trapped ion quantum computation excels is the method of electronic state detection. A subset of the electronic structure of alkali ions, such as $^{40}Ca^+$, have a lambda level structure, which can be used to allow a single quantum absorption event at one frequency to be detected by probing the ion's state with a second resonant frequency. This method has been variously called *electron shelving* [109], a double resonance [110–112], or a quantum counter [113].

In figure 2.13 are the relevant energy levels of $^{40}Ca^+$ used for state detection. For $^{40}Ca^+$, the qubit can be encoded into the $S_{1/2}$ and the $D_{5/2}$ states (often denoted just $|S\rangle$, $|D\rangle$). With reference to figure 2.13, if the ion is illuminated with 397 nm light and the ion is in the $S_{1/2}$ state,

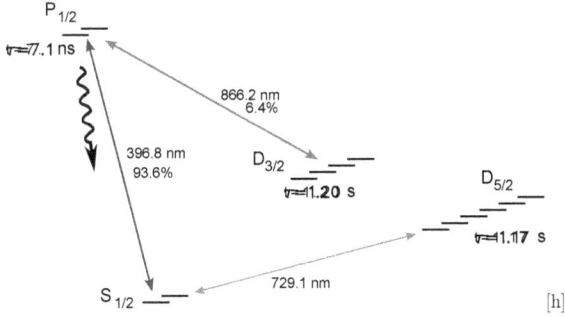

Figure 2.13: State detection using the electron shelving technique for ^{40}Ca$^+$. Absorption of a single 729 nm photon causes an ion in the ground state to be excited to the D$_{5/2}$ state. If the ion is in the D$_{5/2}$ state, it will then cease to fluoresce in the presence of 397 nm radiation. If the qubit is encoded into the S$_{1/2}$ and the D$_{5/2}$ states, then this method can measure the qubit with more than 99.99% accuracy [114]

it will scatter photons. But if it is in the D$_{5/2}$ state, it will remain dark. During state detection, 866 nm light is used to repump out of the D$_{3/2}$ state. Using this method, state detection can be done in 400μs with a fidelity of 99.99% [114]. This performs a canonical projective measurement, so that after measurement, the ion is in one of the two states which encode the qubit. If the state detection is done while driving the qubit, quantum jumps can be seen between the two states of the qubit [115, 116].

Since the measurement is projective, if the measurement basis is not aligned with the state of the ion, then the measurement is probabilistic. For instance, if the ion is in a superposition state $(|S\rangle + |D\rangle)/\sqrt{2}$, then there is a 50% probability of seeing fluorescence when measured. Even if the result of a quantum algorithm expects the ion and measurement to be in the same basis, experimental errors during the execution of the algorithm can cause the ion's state to wander off basis. This means that usually, the experiment is repeated (often 100 times or more) so as to estimate what fraction of the qubit is in $|S\rangle$ or $|D\rangle$.

The measurement basis is described using the Pauli-operators (see eq. 2.63). For instance, the expected value of the measurement $\langle\psi|\sigma_z|\psi\rangle$ represents a measurement in the z basis (i.e. $|S\rangle, |D\rangle$ basis). For a qubit encoded in the $|0\rangle = |D\rangle$ and $|1\rangle = |S\rangle$ states, the expected value is equal to $(P_f \times 2) - 1$, where P_f is the probability of detecting fluorescence. Since probability cannot be measured directly, the experiment must be repeated many times and an estimate of the probability of fluorescence can be inferred. Measurements in the x and y bases cannot be made directly, but can instead be measured by performing a $\theta = \pi/2$ single qubit rotation before measuring in the z basis. For instance, rotating the qubit about the x axis by $\pi/2$ followed by a measurement in the z basis is equivalent to a measurement in the y basis.

2.9   Qubit Decoherence, Dephasing and Decay

Quantum computers and simulators are not unlike analog computers and simulators, in that they essentially use a physical state (the quantum state) directly as a variable for processing. Quantum algorithms, like the Deutsch-Jozsa algorithm [2, 117, 118] are meant to produce an answer in the z-basis of the qubit(s), so that one can obtain a result in a single measurement of whether the qubit is $|0\rangle$ or $|1\rangle$. But while the end result may be measured in one particular basis, small errors during the computation, change the value of the quantum state in unexpected ways. If a system of qubits dephases, loses coherence or decays, the result of any measurement on the qubits may no longer be valid, so that controlling these sources of error is paramount to a functional quantum computer or simulator.

The rest of this section is organized as follows. First an introduction to qubit fidelity is provided (2.9.1). Then an overview of quantum error correction is given (2.9.2). Finally the specific sources of errors are described, first for electronic states of the ion (2.9.3), and then the motional states of the ion (2.9.4).

2.9.1   Qubit Infidelity

Three ways a qubit can be altered are decay, decoherence and dephasing. Decay is when the quantum state, the qubit is encoded in, jumps to another state. This is usually because of spontaneous emission or resonant interaction with random stray fields. Decoherence is when the basis of the qubit rotates in a random way. Considering only pure states and using the Bloch vector notation introduced above in section 2.7.1, decay causes the angle θ to change, and decoherence causes the angle ϕ to change; both in random ways. Dephasing is similar to decoherence, in that the Bloch vector rotates, but it is in-principle reversible; i.e. dephasing is caused by systematic errors in the experimental setup, which if characterized can be corrected for.

Decoherence can also be described using the density matrix notation (see sec. 2.6.1.1). In this case the Bloch vector (see fig. 2.12) is the sum of the probabilistic mixture of pure state vectors. In such a representation, decoherence will shorten the vector in the Bloch sphere. However, for the description of error syndromes below, a pure state description is used.

It is worth mentioning why digital computers do not have this problem. Binary digital machines use non-linear thresholding of a physical state at every step of computation to store and process a binary value. Because for a digital machine, the basis of measurement does not change (its usually a voltage), the non-linear thresholding allows errors to be effectively eliminated at each step of processing.

Attempts to use thresholding techniques on the quantum state are frustrated by two things. First, while a quantum algorithm may produce a final answer in the form of a qubit to be measured either up or down, for many quantum algorithms, the intermediate steps need to use the vast Hilbert space of the qubits to speed up the computation. Secondly, it is not possible to measure the qubit without altering it. Any attempt to threshold it before the algorithm is finished would project it into the unentangled subspace. The Hilbert space of the unentangled qubits is exponentially smaller. The lack of thresholding gives rise to the fear that any small errors could add up at each step in the computation, eventually rendering the result useless.

2.9.2 Quantum Error Correction

Authors such as von Neumann [119], Shannon [120], and Hamming [121] developed a method for averaging out errors in physical states which store or transmit information (also called *channels*) using redundancy. Happily, this can be applied to quantum states in the form of Quantum Error-Correcting Codes (QECC) [103, 122, 123] subject to the condition that the errors are *uncorrelated*. Uncorrelated errors being defined so that N separate qubits, each having an error rate $\epsilon_q << 1$, have the same (or correlated) error likelihood which scales as $(\epsilon_q)^N$.

Fully correlated errors, which are defined here as the error rate a group of N qubits have in excess of the uncorrelated error rate ϵ_q^N are still a source of concern. These sorts of errors can be caused by the qubit laser's frequency drifting or its amplitude fluctuating, or fluctuations of the static magnetic field which provides the Zeeman splitting. Correlated errors can also be caused by ion heating. These errors can cause the processing of the quantum algorithm to be incorrect. Because they can act on more than one qubit at a time, they produce fully correlated errors, which cannot at this time be eliminated with QECC. It remains to be seen, how far these correlated errors can be reduced, to allow for quantum information processing to be scaled-up. For now the reduction of these correlated errors must be done by reducing the sources of error directly, or with quantum engineering methods such as using a decoherence free subspace [124] or by dynamical decoupling [106].

Ion heating is a particular concern for scaling-up ion trap QIP, because it is a correlated source of decoherence and decay who's physical mechanisms are not yet understood. Because of this, an important performance criterion is the ion trap's heating rate. Not all the reasons for ion heating are completely understood, but many mechanisms have been identified [125].

2.9.3 Electronic-Qubit Decoherence

The electronic qubit is subject to decay, dephasing and decoherence. Since, for the $^{40}Ca^+$ ion's optical qubits, the metastable D states have a lifetime of about 1 second, the decay can usually be neglected until the experiment becomes a substantial fraction of this time. Usually what limits the accuracy of experiments with respect to the electronic qubit is its decoherence and dephasing. Two important sources which can limit the ion's coherence time are fluctuating magnetic fields and fluctuations in the driving laser's intensity and frequency.

2.9.3.1 Fluctuating Magnetic Fields

As explained in section 2.2.2, a static magnetic field is applied in the direction of the quantization axis. This gives rise to a Zeeman splitting so that the individual Zeeman sublevels of the electronic states can be addressed. Any fluctuations of the magnetic field in this direction cause the energy-level separation to fluctuate. This can cause the qubit to rotate at a rate other than expected. So that in the rotating frame of the qubit, the phase ϕ is no longer constant. The use of magnetic shielding can reduce this decoherence due to random magnetic fields. If the magnetic field fluctuations have a periodic nature, then synchronizing the experiment with the periodicity, such as done with a line trigger, can also reduce this effect.

2.9.3.2 Laser Frequency and Amplitude Drifts

As described in section 2.7.1, a focussed laser beam can be used to perform single qubit rotations. The speed of this interaction, proportional to the effective Rabi frequency Ω_0, is in turn proportional to the laser's field amplitude (eq. 2.65). As long as the laser fluctuations are small compared to the nominal level, a Taylor expansion gives that the change in the effective Rabi frequency will be proportional to half the change in the laser intensity. Fluctuations in the pointing of the laser beam, motion of the ion, or fluctuations in the power of the laser beam will then cause the effective Rabi frequency Ω_0 to vary proportionally to the level of these fluctuations. Fluctuations of the laser frequency can also cause a change in both the phase of the qubit rotation ϕ, and the effective Rabi frequency.

Since the phase of the qubit relative to the laser must be kept track of, if the qubit laser's frequency drifts unexpectedly, then this will cause errors in subsequent qubit operations. By using ultra-low-expansion glass[2] (ULE™) for the qubit laser's stabilization cavity, the laser can have a drift as low as 60 mHz per second [126], so that this source of error can be effectively eliminated.

Laser intensity variation can however still be a source of error. If the beam-pointing of the laser is unstable, then the ion will experience intensity fluctuations. This can be minimized by using larger beam sizes and more stable optics. For power level fluctuations of the laser beam, intensity stabilization has been used to reduce this error to less than 0.5% [127]. This could be improved upon. For a bandwidth of 1 MHz, laser intensity could possibly be stabilized to within 10 ppm (parts per million) [128, 129]. However, intensity fluctuations can also be caused by undesired motion of the ion relative to the laser.

2.9.3.3 Spectator Modes

The form for the effective Rabi frequency given in section 2.65 is a one dimensional approximation. Motion of the ion(s) in other directions can cause intensity fluctuations, as well as give the ion extra degrees of freedom to move in response to light induced pressure. Since these modes of motion are not considered the main actors in desired light-ion interactions, they are called *spectator modes*.

The full form of the interaction includes the motion from the spectator modes of the ion(s) [77, 102]. The amount of motion in each of the spectator modes is a random distribution

---

[2]Corning Incorporated

after Doppler cooling, and this can cause the effective Rabi frequency to vary from experiment to experiment. If the spectator modes are only Doppler cooled, their motion can cause decoherence, with a time scale of 10s of microseconds, while performing laser mediated qubit gates. For this reason, ground state cooling of all the motional modes is done for high fidelity quantum operations.

Even if the spectator modes are cooled to their motional ground state, they will cause decoherence proportional to the sum of the number of modes times the modes' Lamb-Dicke parameters squared η^2. The Lamb-Dicke parameter is of order 1% for current experiments with laser induced qubit gates. So that the lower limit for single ion operations using lasers is of order 10^{-4}. Since the Lamb-Dicke parameter is inversely proportional to the square-root of the trap frequency ω_x, this source of decoherence could, in principle, be further reduced. However, the number of spectator modes is roughly $3N$, where N is the number of trapped ions. This means that as the number of ions in a trap is increased, the errors due to spectator modes grows linearly. Some proposals for scaling up ion trap quantum computing have advocated placing the ions in separate traps to avoid this. This is covered more in section 2.12.

2.9.4 Motional-Qubit Coherence and Heating

The motional qubit is subject to decay and decoherence. And since the motion of the ions is used to couple quantum information between ions, any heating during this interaction corrupts the quantum logic operation in a correlated way. This means that ion heating could be a serious roadblock to large-scale ion trap QIP.

The charged ion in the trap responds to fluctuating electric fields and these can cause ion heating. Methods to keep the ion cool, by coupling it to a zero temperature bath [130] while preventing the level decay, would not prevent these fields from causing the motional state to lose coherence. So that an important measure of the performance of a particular trap is its heating rate. Another source of error is a drift in the trap frequency, which can cause a dephasing of the ion's motional state.

2.9.4.1 Trap Frequency Stability

The secular trap frequency ω_x, sets the phase progression of the pseudostationary states of the trapped ion. The secular frequency ω_x is proportional to the field gradient parameters α_t, α_s. And these are proportional to the amplitude of the voltages applied to the trap electrodes. If these can be controlled to, at best, a few parts per million (ppm), then the ion's secular frequency will also, at best, also be stable to a few ppm.

For many ion trap experiments, the actual control of the trapping voltage is a few hundred ppm, so that a secular trapping frequency of 1 MHz will vary by a few hundred hertz during experimentation. This limits the coherence time of sideband transitions to approximately 10^{-2} seconds. The voltage fluctuations are primarily caused by experimental errors in the generation of the RF drive. If improved coherence on the motional sidebands is desired, improvement of the voltage stability or active stability of the trap frequency would need to be done.

2.9.4.2 Heating From External Fields

Apart from the large electric potentials which confine the ion, residual fluctuating electric fields in the experiment couple to the motion of the ion and induce transitions between vibrational states. For ion traps with single ions, initially in the ground state, experiments have been constructed where the heating of the ions is less than one phonon per second [125]. However, as the size of the trap is scaled down to microminiature sizes, the heating rate of the ion(s) has increased in many experiments. To understand the possible sources of heating in ion traps requires additional theory, presented here (and in appendix E). There are two main classes of heating, the first of which is linear. The second type of heating grows exponentially in time.

Excitation of the motional state can be caused by a resonance between the trapped ion and a spatially homogeneous field [131, 132]. Spatially homogeneous here means that it does not change appreciably over the range of ion motion present in the experiment. If the exciting field is a coherent source, the amplitude of the motion will grow linear in time (and the energy will then grow as the square of the time). If the exciting field can be described as a white noise spectrum, then the amplitude of the motion will grow with the square root of the time and the energy linearly. For a white noise source, the exciting field is not coherent and this type of excitation is considered heating. A quantitative description of this sort of heating is given below.

The second type of excitation, often called *parametric* excitation, occurs when there is a resonance with a spatially inhomogeneous field [77, 131, 133]. An example of a spatially inhomogeneous field is the quadrupole trapping field from the drive electrodes. The gradient of the field means that as the ion's motion grows, the size of the perturbations also grow. For instance, any voltage fluctuations on the drive electrodes would manifest themselves as an inhomogeneous fluctuating electric field. Parametric heating could also be caused by interactions between a magnetic field and the ion. As the ion's velocity increases with its motion, the Lorentz force would increase. Using laser cooling, trapped atomic ions are kept close to the ground state of their motion. As such, parametric heating processes are not usually considered. But for long storage times with the cooling lasers turned off, parametric heating might be a source of ion heating.

Heating due to homogeneous fluctuating wideband field noise is described mathematically as follows. When the ion is cooled close to the quantum ground state, the corresponding motional dipole moment is $d_I \approx qa_0$, where $a_0 = \sqrt{\hbar/(2m_I\omega_x)} \approx 10\,\mathrm{nm}$ is the ground-state size. This dipole moment makes trapped ions susceptible to electric fields. Adopting, as above, a one-dimensional description of the ion motion along specific axis $\mathbf{e}_t$ of the trap and changing into the interaction picture with respect to the bare trapping Hamiltonian $\hat{H}_t(t)$, the ion-field coupling is given by Leibfried et al. [73] as

$$\hat{H}_{\mathrm{ion-field}}(t) = d_I \left[u(t)\hat{a}^\dagger + u^*(t)\hat{a}\right] \delta E_t(t). \tag{2.91}$$

Here $u(t)$ are solutions of the Mathieu equation 2.13 using the Floquet ansatz (see eq. 2.17). The annihilation and creation operators $(\hat{a}, \hat{a}^\dagger)$ create and destroy the quasistationary states. Here $\delta E_t(t) = E_t(t) - \langle E_t \rangle$, where $E_t(t) = \mathbf{e}_t \cdot \mathbf{E}(\mathbf{r}_I)$, is the fluctuating component of the electric field at the position of the trap, $\mathbf{r}_I$, and along the direction $\mathbf{e}_t$.

Using equation (2.91) and Fermi's golden rule, the heating rate from the ground state to the first excited state Γ_h is given by Wineland et al. [77, 131] as

$$\Gamma_h = \frac{e^2}{4m_l\hbar\omega_x} \sum_{j=-\infty}^{\infty} |C_2 j|^2 S_E((\beta_x/2 + j)\Omega_{rf}), \tag{2.92}$$

where the motional terms $C_0 = 1$ and $C_2 = q_x/4$ are the lowest order approximation to the Mathieu equation (valid when $q_x \ll 1$, see sec. 2.1.2), β_x is the characteristic exponent, and

$$S_E(\omega) = 2 \int_{-\infty}^{\infty} d\tau \langle \delta E_t(\tau) \delta E_t(0) \rangle e^{-i\omega\tau} \tag{2.93}$$

is the single-sided spectral density of the electric-field noise. For homogeneous wideband electric field noise, the heating rate is independent of the motional state of the ion; i.e. it is a constant rate for $S_E(\omega_x)$, m_l and ω_x constant [131]. Considering only the heating due to field noise resonant with the ion's secular frequency ω_x and the first order motional sidebands $\Omega_{rf} \pm \omega_x$, the change in average energy per unit time $\langle \dot{U} \rangle$ is then

$$\langle \dot{U} \rangle = \frac{q^2}{m} \left[S_E(\omega_x) + \frac{q_x^2}{4} S_E(\Omega_{rf} \pm \omega_x) \right] = \langle \dot{n} \rangle \omega_x \hbar, \tag{2.94}$$

where $\langle \dot{n} \rangle$ is the change in the ion's average phonon number per unit time. For field noise which has a substantial gradient, this expression must be modified, especially in the presence of a stray static field (see sec. E.5). Similarly, this expression is not valid if the field noise is narrow band or approximately coherent; i.e. the ion excitation is caused by electric field noise, which has a peak with a spectral width $\delta\omega$ of order (or narrower) than the inverse of the heating time $1/T$. However, most experiment sources of electric field noise are wideband in nature. And the ion can be kept in (or close to) the motional ground state with laser cooling. This allows most experiments to safely ignore parametric heating and narrow band sources of ion excitation.

Methods to measure the heating rate are given below in section 2.10 and in appendix D is a compilation of all known heating rate measurements. In appendix E, a compendium of relevant physical sources of electric-field noise and their characteristics are given.

2.10 Heating Rate Measurements

As described above, the level of heating an ion experiences in any given trap is an important specification of the ion trap experiment. In this section the theoretical basis is given for a few heating rate measurement methods. First a simple method is described, where all sources of cooling are turned off, and the wait time until the ion is lost allows the heating rate to be estimated (2.10.1). A second method uses a resolved sideband method to probe the coherence of the sideband spectroscopy (2.10.2). The last method discussed uses the fact that motion of the ion will diminish the fluorescence of the ion during Doppler cooling (2.10.3).

2.10.1 Ion Loss Heating Rate Measurement

Consider some homogeneous wideband field noise, which can heat the average energy of the ion linearly to above the trap depth as per equation 2.94. One method used to estimate the heating rate of ions in a quadrupole ion trap (QIT) is to turn off any cooling processes and allow the ion to heat up. When an ion's energy exceeds the trap depth, it can be lost. Assuming a linear heating rate, one can then calculate the heating rate as the trap depth Φ_{depth} divided by the uncooled ion lifetime t_{uc}. When the ions have approximately a 50% chance of being lost, then their energy is about equal to the trap depth. Since the ion has three axes, and for each axis, potential or kinetic energy can be stored. The total energy of the ion U is,

$$U = 3k_{\text{B}}T, \tag{2.95}$$

where k_{B} is Boltzmann's constant. For each axis, the average phonon level $\langle n \rangle$ will be

$$\langle n \rangle = \frac{k_{\text{B}}T}{\hbar\omega_j}, \tag{2.96}$$

where ω_j is the frequency of the j-th axis. The time rate of change of the ion's temperature is then approximated as,

$$\dot{T} = \frac{\Phi_{\text{depth}}}{3k_{\text{B}}t_{\text{uc}}}. \tag{2.97}$$

This method to estimate the heating rate can easily be too high, because the assumption that the heating rate is linear is not a very good one as the trap becomes appreciably anharmonic. As the ion becomes very hot, any parametric heating mechanisms could also come into play. As such this method is considered to give an upper bound on the heating rate.

2.10.2 Sideband Heating-Rate Measurement

Besides quantum information processing, one use of laser spectroscopy on long-lived states is to measure the heating rate of the ion. Since an ion heats up in the presence of electric-field noise, this allows one to use ions as probes of the electric field. If the ion is initially in the electronic ground state, with a motional state n, then driving Rabi flops on the BSB will cause the ion to undergo transitions of both the electronic and motional state at a rate given by

$$\Omega_{\text{BSB}} = \Omega_0\sqrt{n+1}\eta \tag{2.98}$$

so that if the ion is initially in the quantum ground state ($n = 0$), then the blue detuned Rabi frequency $\Omega_{\text{BSB}} = \Omega_0\eta$. As the ion is heated (see above in section 2.9.4.2), the motional state increases and becomes a mixture of $n = 0, n = 1, n = 2...$, with associated Rabi frequencies scaled by $\sqrt{1}, \sqrt{2}, \sqrt{3}...$ so that Rabi oscillations driven on the BSB will be composed of multiple frequency components. The superposition of these sine waves will manifest themselves as a loss of contrast on the Rabi oscillations. Modelling the superposition of these various Rabi flops compared to the measured Rabi flops, as the waiting time is increased, can be used to estimate the ion heating rate.

2.10.3 Laser Recooling Heating Rate Measurement

The motion of the ion affects its fluorescence when illuminated by the Doppler cooling beam as explained above in section 2.6.1. One method of measuring the heating rate of the ion uses this effect. By first turning off the cooling light and allowing the ion to heat up, and then turning on the Doppler cooling beam so that the ion is recooled, the fluorescence rate is altered and it is possible to extract the average initial energy $\overline{E}$ of the ion from a fit to the averaged fluorescence rate.

The recooling model used here is one-dimensional and assumes that the ion is in the weak binding regime (trap frequency much less than cooling transition linewidth $\omega_x \leq \gamma$). Equation 2.80 is used to predict the ion's average fluorescence, where the effective detuning δ_{eff}, is caused by the laser detuning as well as the Doppler shift due to the ions thermal motion. The thermal state of the ion is treated as a probabilistic mixture of coherent states [134, 135] so that the average fluorescence can be predicted. For the actual experimental measurement, in order to achieve sufficient statistics, the experiment is repeated many times (typically of order 1000 times). This then gives an expected (from the model) and measured time-dependent fluorescence curve. If the Doppler laser parameters are known, then the only free parameter is the initial average energy (temperature) of the ion in the direction of the cooling laser. Fitting these two curves with different heating times allows the heating rate to be calculated.

2.11 Micromotion

As described above in section 2.1.2, ions in QIT traps have richer dynamics than just that of harmonic oscillators. They also have a motion at the trap drive frequency Ω_{rf}. The rest of this section is organized as follows. First a minimal micromotion is described, which is the case if the ion is oscillating about the RF electric-field null-point of the trap (2.11.1). Next the excess micromotion is given, that an ion has, if the ion is displaced from the trap location (2.11.2). Finally, some methods are described to minimize the micromotion of a trapped ion (2.11.3).

2.11.1 Minimal Micromotion

To find the amplitude of this motion, the solution of the characteristic frequency β_x (eq. 2.20) is substituted into the ansatz (eq. 2.17). When $q_x \ll 1$ the solution can be approximated by taking just the first two terms of the series and setting $A = B$, $c_0 = 1$, $c_2 \simeq q_x/4$. In this case, the x-axis motion of an ion can be approximated as,

$$x(t) \simeq 2A\cos(\beta_x\Omega_{\text{rf}}t/2)[1 - \frac{q_x}{2}\cos(\Omega_{\text{rf}}t)]. \tag{2.99}$$

The motion of the ion then has a harmonic oscillation with amplitude $2A$ at the secular frequency $\omega_x = \beta_x\Omega_{\text{rf}}/2$. Superimposed on top of this secular motion is a faster motion, x_{mm}, given by

$$x_{\text{mm}}(t) = -2A\frac{q_x}{2}\cos(\beta_x\Omega_{\text{rf}}t/2)\cos(\Omega_{\text{rf}}t). \tag{2.100}$$

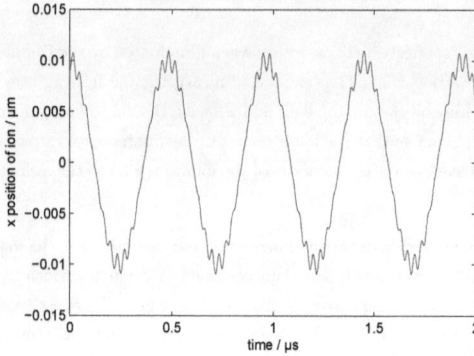

Figure 2.14: An example of the classical motion of an ion in a quadrupole ion trap. The slow oscillations are at the secular frequency of 2 MHz with an amplitude of 10 nm. The fast oscillations are at the trap drive frequency of 30 MHz. The trapping parameter q_x is approximately 0.19, which sets the amplitude of the micromotion relative to the secular motion.

This motion has a frequency of Ω_{rf} with an amplitude scaled by $q_x/2$, relative to the secular motion. Since many experiments operate the trap with $q_x \ll 1$, the amplitude of this motion at the trap drive frequency is small and termed *micromotion*. The further the ion is from the time varying electric-field null (called the *RF null* from here on), the larger are the trapping fields and the associated micromotion. Figure 2.14 shows an example of the trajectory of a trapped ion according to equation 2.99. On top of the secular motion at 2 MHz with an amplitude of 10 nm, is a faster micromotion at 30 MHz with an amplitude of approximately 1 nm.

2.11.2   Excess Micromotion

In experiments, the actual micromotion that the ion experiences can be well in excess of the micromotion expected from the above description. This can be due to undesired stray electric fields. Taking the pseudopotential approximation, the x-component of the static electric field, $E_{\mathrm{S}x}$, will displace the oscillating ion from the center of the trap by an average distance, x_{D} equal to,

$$x_{\mathrm{D}} = \frac{qE_{\mathrm{S}x}}{m_{\mathrm{I}}\omega_{\mathrm{x}}^2}. \tag{2.101}$$

So that the distance the ion is displaced from the center of the trap is inversely proportional to the square of the secular frequency ω_{x}. In such a case, the total motion, using the same approximation as in equation 2.99, is

$$x(t) = A\cos(\beta_x\Omega_{\mathrm{rf}}t/2) + x_{\mathrm{D}} + x_{\mathrm{mm}} - x_{\mathrm{D}}\frac{q_x}{2}\cos(\Omega_{\mathrm{rf}}t). \tag{2.102}$$

The first two terms represent the displaced secular motion, the third term is the normal micromotion (eq. 2.100), and the last term is the excess micromotion x_{emm}. Written in terms of the

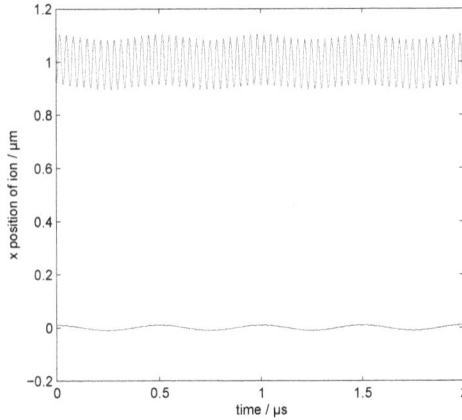

Figure 2.15: The effect of a stray field in an ion trap. The blue trace shows the unperturbed classical trajectory of a trapped ion, with the same parameters as shown in figure 2.14. In the presence of a stray field, the ion's motion is displaced from the RF null (red), and its secular motion is unaltered. However, the excess micromotion is clearly visible.

x-component of the stray electric field E_{Sx}, the excess micromotion has a form of

$$x_{\text{emm}} = q_x \frac{q E_{Sx}}{2 m_1 \omega_x^2} \cos(\Omega_{\text{rf}} t). \tag{2.103}$$

So that the amplitude of the excess micromotion is inversely proportional to the square of the trap frequency ω_x and proportional to the strength of the stray electric field E_{Sx}. Figure 2.15 shows the effects of a stray field on the motion of a trapped ion. For the ion trapped with the same parameters as that given above, a stray field which pushes the ion 1µm from the RF null, causes the micromotion amplitude to increase from 1 nm to 100 nm. Excess micromotion can also be caused by a phase mismatch on a system with multiple RF electrodes. This is described in section 3.5.5. In general, excess micromotion is considered a pathological condition of ion trapping experiments and steps are taken to avoid this condition [136]. However in some cases, it can be desirable; for instance if one would like to use micromotion to address ions in the array [77, 137].

2.11.3 Micromotion Compensation

The presence of DC fields, which push the ion away from the RF null, as explained above, can cause excess micromotion. In order to correct for this, compensation electrodes are sometimes installed in an ion trap experiment. Voltages can be applied to these compensation electrodes so as to null the stray DC field at the ion so that the ion can then fall into the RF null position. Alternatively, the existing electrodes (RF and DC) can have a compensation voltage added to them electronically. The primary challenge to experimenters is to measure the micromotion amplitude, so that the compensation voltages can then be adjusted.

There are several methods to measure the micromotion amplitude or the stray fields, which cause this. One is to measure the position of the ion while lowering the RF trapping field [136]. Another method is to measure the size of the micromotion sidebands using an atomic transition much narrower than the trap drive frequency as explained in 2.4.2. One can also record the time arrivals of the ion's fluorescence photons, so that a lock-in technique can be used [136]. Other ways to measure the micromotion is by parametric excitation of the trapped ions [138] or by using a modulated Raman effect [139].

2.12 Architecture Proposals for Large Scale Integration

As of this writing, single atomic ions trapped in quadrupole ion traps (QIT) are one of the most promising candidates for scaling-up quantum simulation and computation [40]. The criteria, according to DiVincenzo [140], for scaling up quantum computation are shown in table 2.2. While as progress is made in QIP, the bar is raised to satisfy each of these criteria, items 2-5 on this list are well satisfied by ions in a QIT as follows,

2. High fidelity initialization can be done (see sec. 2.4.3)

3. Long motional ($\simeq 1$ s, see sec. 2.9.4) and electronic ($\simeq 10$ ms, see sec. 2.9.3) coherence times compared to the gate time ($\simeq 20$ μs, see sec. 2.7)

4. Mapping between the motional and electronic states of the ions, via spectroscopy with coherent radiation 2.4, allows for a universal set of quantum gates

5. The electronic state can be read-out with the use of the electron shelving method 2.8

6. Cavity QED can convert stationary (electronic) qubits to flying (photonic) qubits

7. An optical fiber can be used to transmit flying (photonic) qubits

What remains is to be shown is whether quantum operations with trapped atomic ions can be sufficiently scaled-up to provide a useful quantum computer. In principle, there seems to be no fundamental reason why trapped atomic ions cannot be scaled-up for large-scale simulation and computation. However, there are many technical reasons which have limited such systems to just tens of qubits.

The rest of this section outlines a state of the art experiment and details some of the proposals to scale-up trapped ion quantum computation and simulation. First a computing architecture based upon a linear ion trap is described (2.12.1). Architectures using multiple ion traps have been suggested as a way to scale-up the system. This would require getting the ions in separate traps to communicate with each other. Several approaches are exist. First a proposed architecture based on a segmented linear trap is given (2.12.2). Next an architecture which transports or shuttles the ions between zones in a segmented linear trap is described (2.12.3).

Criterion for Scalability	in Ion Traps
1. A scalable physical system with well characterized qubits	?
2. The ability to initialize the state of the qubit to a simple fiducial state	✓
3. Long relevant decoherence times, much longer than the gate time	✓
4. A universal set of quantum gates	✓
5. A qubit-specific measurement capability	✓
6. The ability to convert stationary and flying qubits	?
7. The ability to faithfully transmit flying qubits between specified locations	✓

Table 2.2: The DiVincenzo criteria [140] for scaling up quantum computation. Items 2-5 have been shown to be well satisfied by ions in quadrupole ion traps. Items 6 and 7 are part of the necessary requirements to create a quantum network.

A method is then given which uses an optical resonator or cavity to enhance the photonic interaction between ions in different traps (2.12.4). Finally a proposal which uses a lattice (or array) of traps and uses nearest-neighbor interactions is described (2.12.5).

2.12.1 Linear Ion Traps

A state of the art trapped-ion QIP experiment can be built by loading a string of ions into a linear quadrupole ion trap, where each ion can store a single qubit. These qubits can be made to interact via the phonon-bus [40]. However, as the number of ions in the linear trap increases, the number of motional modes (which goes up with 3 times the number of ions) overwhelms the bandwidth of the system.

2.12.2 Linear Arrays

One method to scale-up the system and avoid the issues with having many ions in one trap, is by segmenting the DC electrodes of a linear QIT. Separate trapping zones can be made, so that ions in these trapping zones can be shuttled around or brought to resonance with ions in neighboring zones. Like the coupling described below in section 3.2, nearest neighbor interactions can be made between trapping sites [141, 142]. However, only a one dimensional (1D) array of ion traps can be made using segmented DC electrodes.

2.12.3 Ion Shuttling

A related method to scale up ion traps, is to physically move the ions from storage to processing zones in a segmented linear QIT [143]. One aspect of this proposal, is moving the ions through junctions, where two linear QITs join together. Attempts to send ions through a junction where the time varying field is not modified, so far has resulted in heating the ions [144, 145].

2.12.4   Cavity QED

Ions in different traps, could interact over large distances (even kilometers), with photons (see items 6 and 7 in tab. 2.2). A Fabry-Pérot cavity built around the ion trap could enhance the absorption and spontaneous emission of photons to or from the ion. An optical fiber coupled to this cavity, would then direct the photons to another identical setup. Such a quantum network, has been implemented in atoms [146] and efforts are underway to do the same with ions [147].

2.12.5   2D Arrays

A 2D array of trapped atomic ions would be able to directly simulate a wide variety of 2D quantum phenomena. Proposals have also been given, where a hexagonal 2D array of qubits with nearest-neighbor interactions could be used to directly factor numbers [148]. A rectangular 2D array of qubits with nearest neighbor interactions has also been proposed as part of a scalable quantum computing architecture, where any arbitrary 2D quantum circuit can be mapped to a series of operations on the array [149]. The last proposal is called *measurement* based quantum computing, because after initial preparation of the 2D array in a special state, just measurements need to be made on each qubit individually to created any type of quantum computation. The creation of this special initial state, called a *cluster* state can be done by simulating a nearest neighbor Ising type interaction [150].

In next chapter of this thesis, a 2D array of ion traps, where their nearest neighbor interactions could be controlled by adjusting the position and frequency of the RF null is proposed and analyzed. For instance, this could allow a nearest-neighbor Ising interaction to be synthesized over the whole array via a Trotter approximation [151], so as to create a cluster state for measurement based computing. The rest of this thesis then concentrates on the technical details of how to build and operate such a 2D array of ion traps so that large scale quantum computation and simulation could be practically implemented.

Chapter 3

Addressable 2D Arrays

This chapter describes how to build and operate two dimensional (2D) arrays of ion traps so that one could scale up the QIP techniques described in in chapter 2. This chapter first gives a theoretical basis for the ion dynamics in a 2D array, and then delves into a discussion of the design of a 2D array of addressable ion traps.

The rest of this chapter is organized as follows. First an architecture composed of a 2D array of ion traps is described in section 3.1. Section 3.2 describes a pair of ions, each in a separate trap interacting via a dipole-dipole coupling. Section 3.3 gives details on how this pair of ions could be scaled-up to an interacting 2D array of ion traps. For high-fidelity operation of the 2D array, a method to *address* and/or tune the pairwise interaction of the ions is proposed. Section 3.4 illustrates a away to address the strength of the pairwise interactions between nearest neighbors in a 2D array of ions traps by displacing neighboring ions closer and further apart. A quantitative analysis of RF displacement for a single ion in a trap is given in section 3.5. Next, a discussion of the design considerations of 2D arrays and their scalability is given in section 3.6. Finally, the steps toward a basic 2D algorithm are described in section 3.7.

3.1 Introduction to Addressable 2D Arrays

To create a two dimensional array of qubits, where nearest neighbor quantum logic operations can be performed, it is proposed to use a 2D array of ion traps. Each trap in the array would contain a single trapped ion, which encodes a qubit. For the work here, ^{40}Ca$^+$ is assumed. Using the Coulomb force between nearest neighbors in the array, a neighboring pair of ions can be made to transmit quantum information from one to another. If both ions have the same secular frequency, they are resonantly coupled to each other, and their motional states can be used to perform quantum logic operations or simulate quantum systems. Such a 2D array is depicted in figure 3.1. In the figure is an array of planar-electrode point quadrupole ion traps (see section 2.1.4.1), each which could trap a single atomic ion. Lasers (not shown in figure) could be used to program the ion's electronic and motional states, as well as mediate the interactions between these states (see section 2.4).

The resonantly-coupled Coulomb interaction between two nearest neighbor ions trapped in such an array can be expanded in a power series and the lowest-order relevant term has the mathematical form of a dipole-dipole coupling (described below 3.2). This coherent interaction frequency scales proportionally to the product of the trapped ions' dipole moments μ, and inversely proportional to the separation distance cubed a^3. Considered here are two neighboring trapped ions on resonance with the same mass and charge, so that they have the same secular frequency ω_x and dipole moment μ.

The type of dipole-dipole coupling described here is a near-field effect in that the interaction distance between the ions is typically much smaller than the wavelength of any dipole radiation. Because the interaction is near-field, the energy radiated by the ion-antenna is extremely weak and it can be treated as though no real photon is ever emitted [152]. The transfer of the ion's motion is best described as a near-field coupling which allows the quantized motion of the ion (a phonon) to travel from one ion in a trap to another.

In order to increase the coupling between the ions, either the dipole moment μ can be increased, or the distance a can be decreased. Since the dipole moment squared $|\mu|^2$ is inversely proportional to the secular frequency ω_x, this can be reduced to increase the coupling rate. However, if it is desired to increase the interaction strength by multiple orders of magnitude, it is much more effective to bring the ions closer than to only change their motional frequencies. If one wants to displace the ions in a 2D array without creating excess micromotion, this necessitates modifying the time varying electric fields. One method is described below which uses an adjustable RF voltage applied to a segmented RF electrode placed between each neighboring pair of ion traps in the 2D array. It should be mentioned that the dipole-dipole coupling described below can be in the direction normal to the plane of the 2D array or parallel to it.

Most of the quantum information experiments with trapped ions to date have considered trapping multiple ions in a single harmonic trapping potential [40]. The overall trapping potential of the whole 2D lattice presented here is highly anharmonic. One way to view this, is that anharmonic (pseudo)potentials modify the frequency splitting between the normal modes of motion [70, 153]. The trapping of multiple ions in a single site of the lattice is not treated here, where this would give rise to local modes of motion between the ions in a single site of the lattice. Instead, the relative motion of ions in different sites of the lattice (remote modes of motion) of the lattice is treated here.

While the 2D array of ion traps is physically composed of a lattice of individual point ion traps, the key concept is that of *pairs of neighboring ions*. For this reason the first section in this chapter considers the interaction of just two neighboring trapped ions. These pairs of trapped ions can then be scaled-up into a 2D array.

3.2   Coupled Quantum Harmonic Oscillators

In this section, the dynamics of two ions, in individual harmonic traps, separated by a distance, a, are considered. And methods to use these dynamics to create quantum logic operations are given. First the basis states and the motional mode frequency splitting of two coupled

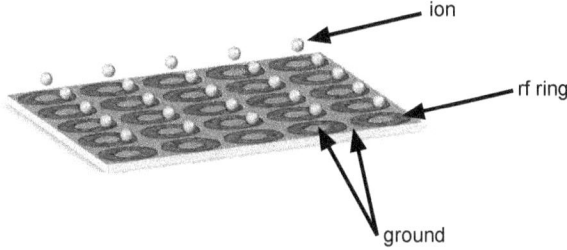

Figure 3.1: A representation of how a 2-dimensional 5×5 array of planar point ion traps might appear. Each ring has a radio-frequency voltage applied to it. Inside each ring is a circular ground electrode. Outside the ring a ground potential can be applied or a distant ground such as the vacuum chamber can be used. The ions are trapped above the circular ground electrodes.

quantum harmonic oscillators are derived (3.2.1). This has the form of a dipole-dipole coupling, where the splitting is calculated for the motional modes parallel and perpendicular to the plane of the 2D array (3.2.2). The approximation of a quadrupole ion trap to a quantum harmonic oscillator is then discussed (3.2.3). Next, the time scale for tuning or addressing this interaction is restricted (3.2.4). This restriction allows for the sudden approximation for projecting between the coupled and uncoupled QHOs bases to be used (3.2.5). Next, using this mathematical framework, a method to create entanglement between the two ions is given (3.2.6). A swap gate is then proposed (3.2.7) and finally a phase gate is described (3.2.8).

3.2.1 Basis States of Coupled QHOs

By using the pseudopotential approximation, the dynamics of two ions, each in separate traps, can be treated like two coupled quantum harmonic oscillators. Quantization of two coupled QHOs proceeds like the uncoupled version with a few extra details. Here, just the coupling of two neighboring harmonic oscillators is presented. However, it turns out that this case is relevant if there is some way to address the interactions in the array [108].

First, just consider the ions' motion in the direction of a line connecting the center of the two harmonic traps. This line is parallel with the x-axis, which is also the same as the x-axis of the point quadrupole ion traps described in the previous chapter. The ions have the same mass m_{I} and electrical charge q. The total classical Hamiltonian can then be written by summing the Hamiltonians from two independent harmonic oscillators (see eq. 2.39) and adding the Coulomb coupling term [108],

$$H = \frac{p_1^2}{2m_{\mathrm{I}}} + \frac{1}{2}m_{\mathrm{I}}\omega_{\mathrm{x}}^2 x_1^2 + \frac{p_2^2}{2m_{\mathrm{I}}} + \frac{1}{2}m_{\mathrm{I}}\omega_{\mathrm{x}}^2 x_2^2 + \frac{q^2}{4\pi\epsilon_0(a_{\mathrm{t}} + (x_2 - x_1))} \tag{3.1}$$

where x_1 and x_2 are the positions of each ion relative to the center of their respective trap, p_1 and p_2 are the momenta respectively, a_{t} is the distance between the RF null positions of the ion traps and ϵ_0 is the permittivity of free space.

We can solve for the motion of the coupled QHO if certain conditions are met. If the motion of the harmonic oscillators is small such that $x_1, x_2 << a_t$, then the coupling term in the Hamiltonian can be approximated to second order as,

$$H \simeq \frac{p_1^2}{2m_I} + \frac{1}{2}m_I\omega_x^2 x_1^2 + \frac{p_2^2}{2m_I} + \frac{1}{2}m_I\omega_x^2 x_2^2 + \frac{q^2}{4\pi\epsilon_0}(1/a_t - x_s/a_t^2 + x_s^2/a_t^3) \qquad (3.2)$$

where $x_s = x_2 - x_1$ is the difference between the positions of each QHO. The zero-th order term $1/a_t$ in the coupling approximation, shifts the absolute level of the Hamiltonian, but does not change the behavior of the system. The first-order term, x_s/a_t^2, has the effect of changing the center of motion of each ion. It pushes them slightly apart to a distance $a \simeq a_t$. The last term however changes the dynamics and is the interesting coupling term.

The Hamiltonian can then be rewritten in terms of the coordinates of two uncoupled motional modes. The modes are called the *stretch* mode and *center-of-mass* (COM) mode. Their coordinates are $x_s = x_2 - x_1$ and $x_c = (x_1 + x_2)/2$ respectively. The form of the Hamiltonian is then,

$$H = \frac{p_c^2}{2m_c} + \frac{1}{2}m_c\omega_x^2 x_c^2 + \frac{p_s^2}{2m_s} + \frac{1}{2}m_s\omega_s^2 x_s^2, \qquad (3.3)$$

where the two new mass terms are defined as, $m_c = 2m_I$, $m_s = m_I/2$, the two new momentum terms are, $p_c = p_1 + p_2$, $p_s = (p_2 - p_1)/2$, and the frequency of the stretch mode ω_s is,

$$\omega_s = \sqrt{\omega_x^2 + \frac{q^2}{\pi\epsilon_0 m_I a^3}}. \qquad (3.4)$$

The double-well trapping potential presented here has substantially modified the relationship between the frequencies of the COM and stretch mode when compared to two ions in a shared harmonic well. That each trap in this double-well potential is itself locally harmonic should not confuse the reader, in that the overall potential is highly anharmonic. Equation 3.3 is separable, so that the Hamiltonian can be written as two uncoupled QHOs,

$$H_c = \frac{p_c^2}{2m_c} + \frac{1}{2}m_c\omega_x^2 x_c^2 \qquad (3.5)$$

$$H_s = \frac{p_s^2}{2m_s} + \frac{1}{2}m_s\omega_s^2 x_s^2 \qquad (3.6)$$

In the so-called *weak-coupling regime*, the frequency difference or splitting $\delta\omega$ between the stretch mode and the center of mass mode is much less than the frequency of the center-of-mass mode ω_x. Using equation 3.4, the frequency difference $\delta\omega = \omega_s - \omega_c$ can be approximated to first order as,

$$\delta\omega = \frac{q^2}{2\pi\epsilon_0 m_I \omega_x a^3} \qquad (3.7)$$

This equation has the form of a dipole-dipole coupling, which can be understood by considering two charged harmonic oscillators near each other as near-field antennas [141].

3.2.2 Dipole-Dipole Coupling

A charged particle in a trap can be considered to form a dipole, where the strength of the dipole $|\boldsymbol{\mu}|$ equals the RMS (root-mean-squared) extent of the particle's motion times its charge q and direction of the dipole is the direction of the particle's motion. The general form of the dipole-dipole interaction energy is E_{dd} [44],

$$E_{\mathrm{dd}} = \frac{|\boldsymbol{\mu}|^2}{4\pi\epsilon_0 a^3}(\cos\theta_{12} - 3\cos\theta_1\cos\theta_2), \tag{3.8}$$

where θ_{12} is the angle between the two axes of each dipole, θ_1, θ_2 are the angle of each dipole with respect to the line connecting their centers. For trapped atomic ions, cooled to the motional ground state, the dipole-moment is given by $|\boldsymbol{\mu}| = a_0 q$, where a_0 is the RMS extent of the ground state wave function, given by $\sqrt{\hbar/(2\omega_{\mathrm{x}} m)}$.

One can use equation 3.8 to calculate the frequency splitting $\delta\omega$ between the x-motional modes of coupled harmonic oscillators. For the example given above of the COM mode and the stretch mode, where the COM mode has $\theta_{12} = 0$, $\theta_1 = 0$, $\theta_2 = 0$ and the stretch mode has $\theta_{12} = \pi$, $\theta_1 = 0$, $\theta_2 = \pi$, the splitting is then,

$$\delta\omega = \frac{\Delta E_{\mathrm{dd}}}{h} = \omega_{\mathrm{s}} - \omega_{\mathrm{x}} \simeq \frac{q^2}{2\pi\epsilon_0 m_1 \omega_{\mathrm{x}} a^3}, \tag{3.9}$$

which is the same result obtained by considering the Coulomb energy of the two motional modes in equation 3.7.

Equation 3.8 can also be used to determine the splitting between modes, who's motion is not parallel to the x-axis. For instance, consider the z-axis rocking and COM motion. These are normal to the line connecting the trap's centers. The splitting between the z-axis COM mode $\theta_{12} = 0$, $\theta_1 = \pi/2$, $\theta_2 = \pi/2$ and the z-axis rocking mode $\theta_{12} = \pi$, $\theta_1 = \pi/2$, $\theta_2 = -\pi/2$ is given by,

$$\delta\omega_r = -\frac{q^2}{4\pi\epsilon_0 m_1 \omega_z a^3}, \tag{3.10}$$

where ω_z is the z-axis motional frequency of the QHOs.

3.2.3 Time Dependent Trapping Potential

There could be some concern that the time dependent nature of a quadrupole ion trap (QIT) would complicate the preceding treatment. This would be the case if the RF trapping fields were different in each trap, either by a shift in the phase or amplitude. This would then cause the quasistationary states to breath with a different amplitude or progress with different reference oscillator frequencies. This would mean that the annihilation and creation operators would not be equivalent for each trap.

To alleviate any concerns about the approximation of a quadrupole ion trap by a harmonic oscillator, certain assumptions are made, which experimentally must then be adhered to. The secular frequency, as well as the phase and amplitude of the trap drive are considered identical for each trap. In the absence of external fields, and provided we restrict ourselves to moments in time that are multiples of $2\pi/\Omega_{rf}$, the ions in the local traps then have exactly the same coupling dynamics as two coupled QHOs.

3.2.4 Coupled to Uncoupled Addressing of QHOs

If it is desired to turn this interaction on and off at will, certain conditions greatly simplify the treatment. Consider a system where the coupling interaction can be turned on much quicker than the splitting $\delta\omega$ (eq. 3.7), but much slower than the trap frequencies ω_x and ω_s, so that the condition is

$$\delta\omega << \omega_a << \omega_x, \omega_s \tag{3.11}$$

where ω_a is the frequency at which the coupling interaction can be turned on and off. If this condition holds, then any change in the coupling interaction is adiabatic with respect to the motional states of each QHO, but sudden with respect to the splitting. This means that when the interaction is turned on, the motional states of each individual QHO are projected onto the coupled basis. This has important implications for creating pairwise entanglement in a controlled fashion without unwanted excitations.

3.2.5 Projecting States between Bases

Consider a system in which the stationary eigenstates are suddenly switched from one set of states to another. The state in a new basis can then be computed by projecting the old state onto the new basis, as per the sudden approximation in quantum mechanics [154]. As described in section 2.3.1, a state of a QHO can be represented in bra/ket notation as,

$$|\psi\rangle = \sum_n c_n |n\rangle \tag{3.12}$$

A product state of the individual QHO states can be written as,

$$|\psi_1\psi_2\rangle = |\psi_1\rangle |\psi_2\rangle = |\psi_1\rangle \otimes |\psi_2\rangle \tag{3.13}$$

where both notations $|\psi_1\rangle |\psi_2\rangle$ and $|\psi_1\psi_2\rangle$ mean that the first QHO has state $|\psi_1\rangle$ and the second QHO has state $|\psi_2\rangle$. In analogy with equation 3.12, consider a set of states $\{\, |j_1\rangle\, \}$ which forms a complete basis for the first QHO, and a set of states $\{\, |k_2\rangle\, \}$ that forms a complete basis for the second QHO. Then any state of the combined system $|\Psi_{1,2}\rangle$ can be expanded as a sum of these product states.

$$|\Psi_{1,2}\rangle = \sum_{j,k} c_{j,k} |j_1 k_2\rangle \,. \tag{3.14}$$

In general there is more than one way to form a complete basis. However for a time independent Hamiltonian, there is at least one basis where each element is a stationary state, allowing the time evolution of the states to be easily solved by using eq. 2.43. This is the case for the coupled and uncoupled QHOs.

For the pair of uncoupled QHOs (see eq. 2.39), the complete set of eigenstates is written as in equation 3.14. The coupled QHOs, since they can be written as separate QHOs in the stretch and COM mode, also have a stationary basis which is formed by the product of the stationary basis for the COM mode $\{\ |n_c\rangle\ \}$ and the stationary basis for the stretch mode $\{\ |m_s\rangle\ \}$,

$$|\Psi_{c,s}\rangle = \sum_{n,m} a_{n,m} |n_c m_s\rangle \qquad (3.15)$$

where $|\Psi_{c,s}\rangle$ is any state of the coupled QHO system, here expressed as a sum of the product states of the individual stationary basis of the COM and stretch mode QHOs. To write the coupled basis in terms of the uncoupled basis, we can write,

$$|n_c\rangle |m_s\rangle = \sum_{j,k} \langle j_1 k_2 | n_c m_s \rangle |j_1 k_2\rangle \qquad (3.16)$$

where we have used the fact that for any complete basis, the identity $I = \sum_{j,k} |jk\rangle \langle jk|$ can be decomposed into that basis. It remains to compute the inner product $\langle j_1 k_1 | n_c m_s \rangle$, which can be done by many means, including by computing the inner product in the position basis using the individual wave functions given by equation 2.41. However, the inner product can be computed using ladder operators (eq. 2.47) as follows

$$\langle j_1 k_2 | n_c m_s \rangle = \langle 0_1 0_2 | \frac{\hat{a}_1^j}{\sqrt{(j+1)!}} \frac{\hat{a}_2^k}{\sqrt{(k+1)!}} \frac{(\hat{a}_c^\dagger)^n}{\sqrt{(n+1)!}} \frac{(\hat{a}_s^\dagger)^m}{\sqrt{(m+1)!}} |0_c 0_s\rangle \qquad (3.17)$$

$$\langle 0_1 0_2 | 0_c 0_s \rangle = 1$$

where the relationship between the ladder operators can be derived using the relations between the position and momentum, and their operators for the QHOs. The ladder operators for the COM and stretch modes written terms of the uncoupled bases' ladder operators are given as

$$\hat{a}_c = \frac{1}{\sqrt{2}}(\hat{a}_1 + \hat{a}_2) \qquad (3.18)$$

$$\hat{a}_c^\dagger = \frac{1}{\sqrt{2}}(\hat{a}_1^\dagger + \hat{a}_2^\dagger)$$

$$\hat{a}_s = \frac{1}{\sqrt{2}}\sqrt{\frac{\omega_s}{\omega_x}}(\hat{a}_2 - \hat{a}_1)$$

$$\hat{a}_s^\dagger = \frac{1}{\sqrt{2}}\sqrt{\frac{\omega_s}{\omega_x}}(\hat{a}_2^\dagger - \hat{a}_1^\dagger).$$

3.2.6   Creating Entanglement with Controlled Coupling

As a simple example, two initially uncoupled QHOs, where one phonon is in the first QHO and the other is in the ground state. Evaluation of equation 3.17 leads to

$$\langle 1_0 2 | n_c m_s \rangle = \langle 0_1 0_2 | \hat{a}_1 \frac{(\hat{a}_c^\dagger)^n}{\sqrt{(n+1)!}} \frac{(\hat{a}_s^\dagger)^m}{\sqrt{(m+1)!}} |0_c 0_s\rangle \qquad (3.19)$$

so that using the relations of equations 3.18 allows one to write the product state in the uncoupled basis as a coherent sum of product states in the coupled basis. Taking the approximation (weak-coupling regime) that $\sqrt{\omega_s/\omega_x} \simeq 1$, a single phonon in the first QHO with the other QHO in the ground state, can be written in the coupled basis as,

$$|1_1\rangle |0_2\rangle = \frac{\sqrt{2}}{2}(|0_c\rangle |1_s\rangle + |1_c\rangle |0_s\rangle). \tag{3.20}$$

However, because of the small frequency splitting $\delta\omega$ between the COM and the stretch mode, the phase of these two modes will propagate at different speeds as a function of time in the same way as equation 2.43. By turning the interaction off at the right time, entanglement can then be created in the uncoupled basis. If the coupling interaction is on long enough to allow for the stretch mode to accumulate a phase of $\pi/2$ relative to the COM, then the above state will evolve to,

$$\frac{\sqrt{2}}{2}\left\{e^{-i\pi/2}|0_c\rangle |1_s\rangle + |1_c\rangle |0_s\rangle\right\} = \frac{\sqrt{2}}{2}\left\{|0_1 1_2\rangle + i\,|1_1 0_2\rangle\right\}. \tag{3.21}$$

Entanglement remains in the uncoupled basis after the interaction has been switched off. The required time to create the entanglement described here is the time $T_{\pi/2}$ for the coupled QHOs to accumulate a phase of $\pi/2$,

$$T_{\pi/2} = \frac{2\pi^2 \epsilon_0 m\omega_x a^3}{q^2} \tag{3.22}$$

Here equation 3.7 has been used to determine the frequency splitting. This method entangles the motional states. Using side-band resolved spectroscopy, the entanglement can be then swapped to the electronic states of neighboring ions in a 2D array as described in section 2.4.

3.2.7   Swap Gate

Although some 2D architecture proposals, and especially 2D simulations, would only need nearest neighbor interactions. It could be desirable to exchange information further out into the 2D array. A conceptually simple way to do this is to perform a swap operation. By turning on the coupling between nearest neighbors, one needs to wait the amount of time that it takes for a phonon to completely move over between neighbors, or a π gate. In this way, a quantum state, for instance entanglement, could be extended further out into the array. The time for a swap gate using the dipole-dipole coupling would be,

$$T_\pi = \frac{4\pi^2 \epsilon_0 m\omega_x a^3}{q^2}. \tag{3.23}$$

3.2.8   Nearest Neighbor Gate Operations

Creating entanglement as described above creates a resource, which can be used for quantum computation such as measurement based computing (see sec. 2.12.5). However, if one would like to create a quantum logic gate, such as a controlled phase gate, another series of steps must be taken. There are a number of proposed physical interactions which could produce a quantum logic gate between trapped atomic ions. One such scheme uses a Mølmer-Sørensen [155] type interaction to perform a controlled phase gate. Another type of interaction with a rapidly rotating magnetic field has improved noise immunity [106].

One detail that must be dealt with is that most of the above proposed interactions have been developed for experiments with multiple ions in a single ion trap. In such a case, the relative splitting between the COM and stretch motional modes is fixed at $\sqrt{3}$, due to the physics of harmonic traps [40]. Whereas the splitting between the COM and stretch modes of motion in 2D arrays of traps is governed by equation 3.4. For 2D arrays where the inter-trap spacing a is of order 100µm, this means that the COM and stretch modes are only separated by approximately 1-10 kHz instead of the approximately 1 MHz for multiple ions in the same trap. This could pose a difficulty if one wants to only address only one of the motional sidebands (COM or stretch).

Gates using an effective spin-spin interaction between ions in separate wells have been able to generate entanglement in their internal electronic states [156]. But for circuit based quantum information processing, one proposal is to use a Cirac-Zoller type controlled-phase gate [104]. Such a proposal [157] to use nearest neighbor dipole-dipole coupling to create a quantum logic gate is as follows: One could start with the ions cooled to the quantum ground state. If the first ion, called the control ion, has an excited electronic state, then a RSB pulse of length π will place the ion into motion if it is electronically excited. The dipole-dipole coupling can then be used to move the motion over to the second ion, called the target ion. A RSB pulse of length 2π will cause the sign of the excited state of the target ion to flip (see section 2.7.1). The final steps are to allow the motion to return to the control ion, and then map the motion back to the control ion's electronic state. For 2D arrays with a nearest neighbor separation larger than about 10µm, the dipole-dipole coupling will be the limiting step in this process. For this quantum logic operation, the gate time is then approximately twice the time for the motional entanglement given above in equation 3.22 and is given by,

$$T_{gate} = \frac{4\pi^2 \epsilon_0 m \omega_x a^3}{q^2}. \tag{3.24}$$

Another proposed controlled-phase gate for ions in separate traps, given by Cirac and Zoller [108], uses a highly detuned laser to produce a state dependent force which is equal to that which displaces the ion the size of the trapped ion's motional ground-state. With this relatively weak state-dependent force, the gate is operated in a so-called *stiff-mode* [158] regime, where the interaction strength has a dipolar decay law. The time for this controlled-phase gate is exactly the same as that given in equation 3.24.

For the above proposed quantum logic gates, the dipole-dipole interaction is the rate limiting step. By reducing the distance between the ions, the time for an entangling gate scales as the inverse of the distance cubed. Also by reducing the trap frequency, the gate time can be reduced proportionally. If gate operations utilizing stronger state dependent forces could be employed [159] then the gate times could also be significantly reduced. However, for the analysis presented here, it is assumed that only gate operations employing relatively weak state-dependent forces, which only change the motional Fock state by 1, are used.

3.3 Scaling-Up Two Neighbors to a 2D Array

If a single ion, trapped in its own trap is considered as a unit, these units can be repeated in a two dimensional lattice. The system of trapped ions could then be scaled up with a 2D lattice of ion traps. An example of a planar electrode structure capable of trapping a 2D array of ions is depicted in figure 3.1. It shows an array of planar electrode ring traps [160]. As the size of the array is shrunk, the dipole-dipole interaction between the ions will increase.

In principle, one can use the same mathematical techniques given above in section 3.2, albeit with more degrees-of-freedom, to predict the dynamics of the whole array. In order to build a functional quantum computer or simulator, it would be necessary to have control over both the individual ions as well as the strength of the interactions between the ions in the array. This requires a method to tune and/or *address* (turn on and off) the interactions between the ions in the array. One possible way to operate a 2D array, would be to only turn on pairwise interactions and perform interactions between nearest neighbors. This would allow for a high degree of control over the interactions between the qubits encoded onto the ions in the array. This places the focus on the interaction between pairs of ions, just like the two coupled QHOs discussed in section 3.2.

In section 3.2, the dynamics of two neighboring traps with the same frequency ω_x was studied. There are many ways to address or tune the coupling, and some of them are discussed below in 3.6.4. However here, a method is considered which uses ion displacement as well as shared frequency tuning. All the ions in an array are initially assumed to be in a state such that the interactions between ions are sufficiently weak that they can be ignored. If a neighboring pair of ions has the same trap frequency ω_x, equation 3.7 gives that the splitting $\delta\omega$ between the COM and stretch modes of the motion is inversely proportional to the trap frequency ω_x and the cube of the ion-ion separation a. So that to turn on the interaction (increase the mode splitting $\delta\omega$) between nearest neighbors, the distance between the ions a and/or their trap frequency ω_x needs to be reduced.

3.4 RF Addressable 2D Arrays

In a point ion trap (see sec. 2.1.1), the location of the RF electic-field null-point (RF null) defines the micromotion-minimized trapping location of the ions. In order to modify the location of the RF null, DC electric fields cannot be used. Instead, if the RF electrodes can be segmented and the applied RF voltage adjusted, the location of the RF null, and hence the minimal micromotion ion location, can be modified. The gradients of the electric field are also modified by this voltage adjustment, so that the trapping frequency also changes. Such an adjustable RF architecture would allow for an array of point ion traps to be addressed and tuned. This RF addressing would allow for adjustment of the location of the RF null (ion location) and the trapping frequencies of two neighboring ion traps.

The rest of this section is organized as follows. First, a quartic model of a double-well potential is described (3.4.1). Next, the dynamics of the ions in two trapping regimes, in this double-well potential, are given (3.4.2 and 3.4.3). Then an RF electrode structure capable of producing double-well pseudopotential is described (3.4.4). The use of RF voltages to produce a double-well pseudopotential gives rise to a coupling between the applied RF voltages and the potential terms. However, the qualitative behavior of the double-well as a single adjustable voltage is varied is shown to remain the same. Finally, it is described how an array of point quadrupole ion traps can be designed with an adjustable RF electrode between them, so as to scale up a 2D array of addressable point ion traps (3.4.5).

3.4.1 Quartic Double-Well Potential

The basic building block of the proposed architecture is that of the double-well potential. In section 3.2, the interaction of two coupled quantum harmonic oscillators was determined. Here a potential model, which approximates this system is given. A mathematical model of a double-well potential, in the x-direction, using quadratic and quartic terms [69, 161] is given by

$$\phi = (a_2 + m_2)x^2 + (a_4 + m_4)x^4, \tag{3.25}$$

where a_2 is the strength of an adjustable quadratic trapping potential, a_4 is the strength of an adjustable quartic trapping potential, m_2 is the strength of a positive static quadratic trapping potential and m_4 is the strength of a positive static quartic trapping potential.

3.4.2 Two Weakly Coupled Ions in an Adjustable Quartic Potential

Consider an ion trap electrode structure, designed and operated in such a way, that its pseudopotential could be approximated by this model. If the quadratic term a_2 could be varied by tuning an RF voltage amplitude, then ions could be trapped, with a tunable potential minimum separation a_t and trapping frequency ω_x. When the anti-trapping term $a_2 < 0$ is strong enough ($a_2 + m_2 < 0$), the positions of the RF null at $x = \pm x_t$ can be solved for as

$$x_t = \sqrt{\frac{|a_2 + m_2|}{2(a_4 + m_4)}} = a_t/2, \tag{3.26}$$

where a_t is the distance between the electric-field null points of the trap. The potential energy U_1 of a single ion trapped in the left well can then be written in terms of its displacement from the RF null position x_1,

$$U_1 = q \left[(a_2 + m_2)(x_1 - x_t)^2 + (a_4 + m_4)(x_1 - x_t)^4 \right], \tag{3.27}$$

where $x = x_1 - x_t$. If the displacement of the ion from the RF null x_1 is much smaller than the distance between the traps $a_t = 2x_t$, then the higher order cubic and quartic terms in U_1 will not have much effect on the harmonic motion around the RF null. For a single ion in the left well with harmonic motion, the potential energy U_1 can then be approximated with just the quadratic terms, which are given by,

$$U_1 \simeq q \left[(a_2 + m_2)x_1^2 + 6(a_4 + m_4)x_1^2 x_t^2 \right].$$ (3.28)

The trapping frequency of a single ion in either well ω_x is then found by equating the potential energy with that of a harmonic oscillator $U_{HO} = m_I \omega_x^2 x_1^2 / 2$. The trapping frequency is then

$$\omega_x = \sqrt{\frac{4q|a_2 + m_2|}{m_I}},$$ (3.29)

where q is the charge of the ion and m_I is its mass. This equation is valid as long as the displacement of the ion from the potential minimum is much smaller than the position of the potential minimum x_{null} ($x_1 \ll |a_2 + m_2|/(a_4 + m_4)$). If two ions are trapped, one in each well, with this condition, then they are said to be *weakly coupled*. In this situation, the dynamics of two ions can be described with a potential minima separation a_t and frequency ω_x as is done in section 3.2. This gives rise to a coupling with a dipole-dipole form. The dynamics of two ions, when the coupling is not weak, is considered next.

3.4.3 Two Strongly Coupled Ions in an Adjustable Quartic Potential

When two ions are trapped, one in each well and the overall trapping potential is tuned from a double-well to a single well, the term $a_2 + m_2$ will at some point be zero so that the quartic term, as well as the Coulomb interaction energy must be taken into account to describe the dynamics of the ions. In this case, the equilibrium distance a between the ions cannot be approximated as the distance between the trapping potential minima a_t because of the ion's Coulomb repulsion. The distance between the ions a including the Coulomb potential is described by the quintic equation

$$(a_2 + m_2)a^3 + (a_4 + m_4)\frac{a^5}{2} = \frac{q}{4\pi\epsilon_0},$$ (3.30)

where ϵ_0 is the permittivity of free space. Such equations do not have simple closed-form solutions, however they can be solved numerically.

Figure 3.2 shows the trapping location of ions in such a double-well potential ϕ as a function of position x for various strengths of the adjustable anti-trapping terms a_2 and a_4, where the adjustable quartic term is proportional to the adjustable quadratic term as $a_4 = a_2 \times 10^5/\mathrm{mm}^2$. The fixed terms m_2 and m_4 are kept constant at $m_2 = 5\,\mathrm{V/mm}^2$ and $m_4 = 50\,\mathrm{V/mm}^4$. The quartic term $m_4 x^4$, which dominates at large distances, forms an overall potential well. The remaining quadratic terms, which dominate at small distances, can form an anti-trapping potential at the center of the double-well. Circles show the location of trapped ions for various values of a_2, a_4. The distance between the potential minima changes as the strength of the adjustable terms a_2 and a_4 are varied. When $a_2 + m_2 \geq 0$, then the two traps have merged into a single trap. When $a_2 + m_2 = 0$, their is no longer a quadrupole trap and the quartic terms a_4, m_4 dominate. If the trapped particles are singly charged ions in such a potential, then their Coulomb repulsion pushes them away from each other as is shown for the case when $a_2 = -5\,\mathrm{V/mm}^2$ (see figure 3.2).

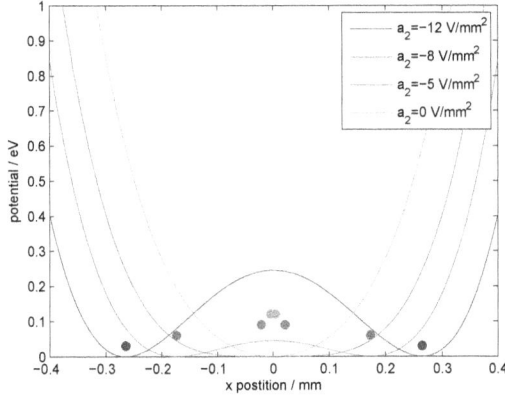

Figure 3.2: The double-well potential ϕ as a function of position x for various strengths of the adjustable anti-trapping terms a_2 and a_4, where the adjustable quartic term scales with a_2 as $a_4 = a_2 \times 10^5$ mm^2. The fixed terms m_2 and m_4 are kept constant at $m_2 = 5$ V/mm^2 and $m_4 = 50$ V/mm^4. The quartic term $m_4 x^4$, which dominates at large distances, forms an overall trapping potential well. The remaining quadratic terms, which dominate at small distances, can form an anti-trapping potential at the center of the double-well. Circles show the location of trapped ions for various values of a_2.

Once the equilibrium distance between the ions a is know, the curvature of the potentials at the equilibrium positions can be calculated. If the trap is operated between the ground state and the first excited state, then the quartic potential can be approximately treated as a quadratic potential. The center-of-mass mode frequency ω_x and the stretch mode frequency ω_s can then be computed as,

$$\omega_x^2 = \frac{q}{m_1}\left[2(a_2 + m_2) + 3(a_4 + m_4)a^2\right]$$
$$\omega_s^2 = \omega_x^2 + \frac{q^2}{\pi\epsilon_0 m_1 a^3}. \tag{3.31}$$

If the splitting $\delta\omega = \omega_x - \omega_s$ is small compared to the motional frequency ($\delta\omega/\omega_x \ll 1$), then the coupling is weak and the splitting can be approximated as

$$\delta\omega = \frac{q^2}{2\pi\epsilon_0 m_1 a^3 \omega_x}, \tag{3.32}$$

which is the same dipole-dipole coupling as equation 3.7. If the coupling is strong so that there is only one trap and it is operated so that the quartic terms are small compared to the quadrupole terms ($x_1 \ll [(a_2 + m_2)/m_4]$), then the equilibrium distance between the ions a is given by

$$a^3 = \frac{q}{4\pi\epsilon_0(a_2 + m_2)}, \tag{3.33}$$

and the stretch mode is related to the center-of-mass mode simply as $\omega_s = \omega_x\sqrt{3}$. This is the situation when there is a well defined single harmonic potential as described above in section 2.3.5.

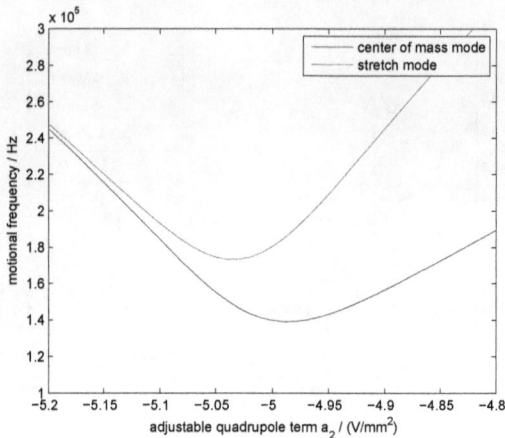

Figure 3.3: The motional frequencies during the merging of a double-well formed from an adjustable quadratic term a_2 and an overall quartic term.

In between the two extremes of weak-coupling and a single harmonic trap, the ions are confined in a strongly quartic potential. The ground state frequencies can be computed from equation 3.31. Figure 3.3 shows the center-of-mass and stretch mode frequencies of the ground states of two ions in a combined quadratic-quartic potential as it is tuned from a double-well to a single-well pseudopotential. Here the behavior is shown around $a_2 + m_2 = 0, a_4 = 0, m_2 = 5 \, \mathrm{V/mm}^2$ and $m_4 = 50 \, \mathrm{V/mm}^2$. As the trap is tuned from a double-well potential to a single well potential, the relative splitting between the center-of-mass mode and the stretch mode changes considerably. On the left side of the figure, the relative splitting of the two modes is weak and has a dipole-dipole form. And on the right side of the figure, the relative splitting is strong and equal to $\sqrt{3}$. In between the Coulomb repulsion pushes the ions away from each other far enough up onto the quartic potential, so that there still is an appreciable curvature of the potential and corresponding lowest energy-level trap frequency ω_x. Next an electrode structure and a method of driving it, which provides a similar adjustable double-well pseudopotential, is described.

3.4.4 Adjustable RF Quartic Pseudopotential

The electric field and the resulting quartic pseudopotential formed from an electrode configuration of two neighboring surface-electrode point ion traps with an adjustable electrode is described here. A double-well pseudopotential along a line connecting two neighboring ions, each in a point ion trap, is shown in figure 3.4. There are three types of electrodes in this configuration. The first are ground electrodes, which have no RF voltage applied. The second are the main RF electrodes, which have a fixed amplitude RF voltage applied to them. The third kind of electrode is an adjustable RF electrode. Figure 3.4a depicts the double-well pseudopotential ϕ of an array when all RF electrodes have the same amplitude. And figure 3.4b shows the pseudopotential ϕ when an adjustable RF electrode, in between the two trapping sites, has its

RF voltage amplitude reduced. When the adjustable voltage amplitude is decreased, the electric field between the two trapping sites decreases as described below. The pseudopotential then decreases as the electric field squared (see eq. 2.25).

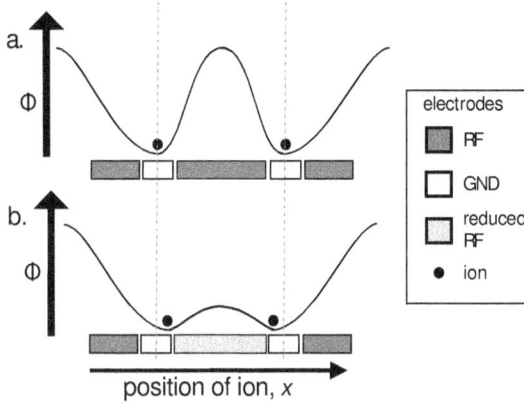

Figure 3.4: A double-well pseudopotential along a line connecting two neighboring ions, each in a point ion trap with a shared adjustable RF electrode. There are three types of electrodes in this configuration. The first are ground electrodes, which have no RF voltage applied. The second are the main RF electrodes, which have a fixed amplitude RF voltage applied to them. The third kind of electrode is an adjustable RF electrode. Figure 3.4a depicts the double-well pseudopotential ϕ of an array when all RF electrodes have the same amplitude. And figure 3.4b shows the pseudopotential ϕ when an adjustable RF electrode, in between the two trapping sites, has its RF voltage amplitude reduced. When the adjustable voltage amplitude is decreased, the electric field between the two trapping sites decreases as described below. The pseudopotential then decreases as the electric field squared (see eq. 2.25).

One method of analyzing the electric field and the resulting pseudopotential then produced by RF voltages on these electrodes is to expand the electric field in a power series. For instance the x-component of the electric field $E_{x,\mathrm{main}}$ (with y, z fixed at $y = 0$ and $z = z_t$, where z_t is the height of the RF null) generated by a voltage, V_{nom} on the main outer RF electrodes can be written as

$$E_{x,\mathrm{main}} = V_{\mathrm{nom}} \left[m_{e,1} + m_{e,2}x + m_{e,3}x^2 + m_{e,4}x^3 + ... \right], \qquad (3.34)$$

where the terms $m_{e,n}$ are called the electric-field moments of the electrode. For a symmetric electrode structure like that shown in figure 3.4, the odd numbered electrode moments $m_{e,1}, m_{e,3}, ...$ are zero. If one is concerned with a sufficiently small $|x|$, then the higher order moments can be ignored. Considering just the first two non-zero terms of this expansion, the outer main RF electrodes then have non-zero moments $m_{e,2}$ and $m_{e,4}$.

The electric field due to an adjustable voltage V_{adj} on the inner electrode can be similarly expanded, where it has moments $a_{e,2}$ and $a_{e,4}$. The total electric field in the x-direction can then be approximated as

$$E_x \simeq (V_{\mathrm{nom}}m_{e,2} + V_{\mathrm{adj}}a_{e,2})x + (V_{\mathrm{nom}}m_{e,4} + V_{\mathrm{adj}}a_{e,4})x^3. \qquad (3.35)$$

The location of the electric-field null x_t in the x direction can be found and is the same as that for the quartic electrical potential above. Written in terms of the electrode moments it is

$$x_t = \pm\sqrt{\frac{V_r a_{e,2} + m_{e,2}}{V_r a_{e,4} + m_{e,4}}}, \tag{3.36}$$

where V_r is the ratio between the two voltages $V_r = V_{\text{adj}}/V_{\text{nom}}$.

If the quartic potential due to static DC fields were to be calculated, then this electric field could be integrated along x to give a the potential. However, for 2D arrays of point ion traps, the pseudopotential is adjusted by changing the amplitude of the time varying electric field, which results in a change of the pseudopotential. The computation of the pseudopotential as a function of the time varying voltages V_{nom} and V_{adj} is done as follows.

The pseudopotential (eq. 2.25 is proportional to the square of the norm of the electric field $|E_x|^2 + |E_y|^2 + |E_z|^2$. Considering first just the x-component of the electric field, this can be written in terms of the various electric-field moments and voltages, described above, as

$$|E_x|^2 \simeq V_{\text{nom}}^2 \{ (m_{e,2}^2 + V_r^2 a_{e,2}^2 + 2V_r m_{e,2} a_{e,2}) x^2 + $$
$$[m_{e,2} m_{e,4} + V_r(a_{e,2} m_{e,4} + m_{e,2} a_{e,4}) + V_r^2 a_{e,2} a_{e,4}] \, x^4 \}, \tag{3.37}$$

where the sextic (x^6) and higher order terms have been ignored. The y-component of the electric field is assumed to be zero due to symmetry of the given geometry. However, the full form of the pseudopotential would also include terms from the z component of the electric field $|E_z|^2$. If the electric-field moments are found by using a simulation of a particular electrode structure, then these can be found directly. However, Laplace's equation 2.1 places some restrictions on the relationship between the moments in the different directions. The electric field in the z direction can be related to the spatial derivative of the electric field in the other directions as

$$E_z = K_z - z \left(\frac{\partial E_x}{\partial x} + \frac{\partial E_y}{\partial y} \right)$$
$$E_z = K_z - z \left\{ V_{\text{nom}} \left[(m_{e,2} + V_r a_{e,2}) + 3(m_{e,4} + V_r a_{e,4}) x^2 \right] + \frac{\partial E_y}{\partial y} \right\}, \tag{3.38}$$

where K_z is a constant of integration, which can be found by setting the total electric field to zero at the RF Null location ($\boldsymbol{E}(x = x_t, y = 0, z = z_t) = 0$), and z_t is the height of the RF null above the trap. Realistic values of the electrode moments as well as the height of the RF null above the electrode surface z_t can be found by simulating an ion trap electrode structure with finite element analysis programs or by using experimentally measured ion frequencies and trap heights.

The pseudopotential is equal proportional to the electric field squared as given in equation 2.25 as

$$\phi_{\text{s}}(\boldsymbol{r}) = \frac{q^2 |\boldsymbol{E}(\boldsymbol{r})|^2}{4m\Omega_{\text{rf}}^2}. \tag{3.39}$$

Using the multipole expansion of the electric field, the pseudopotential along a line in the x direction, which connects the two RF null locations, can then be written as

$$
\phi_\mathrm{s} = \frac{q^2}{4m\Omega_\mathrm{rf}^2} \left\{ K_z^2 - 2z_t K_z \left[V_\mathrm{nom}(m_{\mathrm{e},2} + V_r a_{\mathrm{e},2} + 3x^2 m_{\mathrm{e},4} + 3x^2 V_r a_{\mathrm{e},4}) + \frac{\partial E_y}{\partial y} \right] + \right.
$$
$$
V_\mathrm{nom} z_t^2 \left[V_\mathrm{nom}(m_{\mathrm{e},2}^2 + 2V_r a_{\mathrm{e},2} m_{\mathrm{e},2} + V_r^2 a_{\mathrm{e},2} a_{\mathrm{e},2}) + 2\frac{\partial E_y}{\partial y}(m_{\mathrm{e},2} + V_r a_{\mathrm{e},2}) \right] +
$$
$$
6x^2 V_\mathrm{nom} z_t^2 \left[V_\mathrm{nom}(m_{\mathrm{e},2} m_{\mathrm{e},4} + V_r m_{\mathrm{e},2} a_{\mathrm{e},4} + V_r m_{\mathrm{e},4} a_{\mathrm{e},2} + V_r^2 a_{\mathrm{e},2} a_{\mathrm{e},4}) + \frac{\partial E_y}{\partial y}(m_{\mathrm{e},4} + V_r a_{\mathrm{e},4}) \right] +
$$
$$
9x^4 V_\mathrm{nom}^2 z_t^2 \left[m_{\mathrm{e},4}^2 + 2V_r a_{\mathrm{e},4} m_{\mathrm{e},4} + V_r^2 a_{\mathrm{e},4}^2 \right] +
$$
$$
\left. z_t^2 \left[\frac{\partial E_y}{\partial y} \right]^2 \right\} \tag{3.40}
$$

where z_t is the height of the RF null above the trap. By collecting the quadratic and quartic terms (in x), the terms of the quartic double-well model in equation 3.25 can then be related to the RF voltages and the electrode moments as

$$
m_2 = \frac{qV_\mathrm{nom}}{4m\Omega_\mathrm{rf}^2} \left\{ 6z_t^2 \left[V_\mathrm{nom} m_{\mathrm{e},2} m_{\mathrm{e},4} + \frac{\partial E_y}{\partial y} m_{\mathrm{e},4} \right] - 6z_t K_z m_{\mathrm{e},4} \right\}
$$
$$
m_4 = \frac{qV_\mathrm{nom}^2}{4m\Omega_\mathrm{rf}^2} 9z_t^2 m_{\mathrm{e},4}^2
$$
$$
a_2 = \frac{qV_\mathrm{nom}}{4m\Omega_\mathrm{rf}^2} \left\{ 6z_t^2 \left[V_\mathrm{nom}(V_r m_{\mathrm{e},2} a_{\mathrm{e},4} + V_r m_{\mathrm{e},4} a_{\mathrm{e},2} + V_r^2 a_{\mathrm{e},2} a_{\mathrm{e},4}) + \frac{\partial E_y}{\partial y} V_r a_{\mathrm{e},4} \right] - 6z_t K_z V_r a_{\mathrm{e},4} \right\}
$$
$$
a_4 = \frac{qV_\mathrm{nom}^2}{4m\Omega_\mathrm{rf}^2} 9z_t^2 \left[2V_r a_{\mathrm{e},4} m_{\mathrm{e},4} + V_r^2 a_{\mathrm{e},4}^2 \right] . \tag{3.41}
$$

The terms which are constant in x can be ignored as they must sum to zero, in order for the electric field and hence the pseudopotential to be zero at the trapping location.

Using the values in equation 3.41 for the pseudopotential terms in equation 3.25, figure 3.5 shows the calculated pseudopotential along the x-direction for an electrode structure similar to that shown in figure 3.4, as the voltage ratio $V_r = V_\mathrm{adj}/V_\mathrm{nom}$ is varied from zero to one. When V_r is zero, there is one well defined trap, as the V_r is increased, two traps appear. When $V_r = 1$ there are two well defined traps. For this simulation, $V_\mathrm{nom} = 100\,\mathrm{V}$ and the electrode moments used were $m_{\mathrm{e},2} = 0.7\,/\mathrm{mm}^2$, $a_{\mathrm{e},2} = -2\,/\mathrm{mm}^2$, $m_{\mathrm{e},4} = 5\,/\mathrm{mm}^4$, and $a_{\mathrm{e},4} = 0.5\,/\mathrm{mm}^4$. And the curvature of the electric field in the y direction $\partial E_y/\partial y$ was proportional to the curvature in the x direction plus an offset equal to half the curvature in the x direction at $V_r = 1$. This was justified physically, because in point ion traps traps, the ion is trapped in all three directions. And for the geometry shown in figure 3.4, the curvature of the electric field in the x and y direction are the same when $V_r = 1$.

The coupling introduced by the form of the pseudopotential makes the form of the quartic and quadrupole terms in equations 3.41 complicated. However the behavior of the tuned quartic double-well pseudopotential in figure 3.5 is qualitatively similar to the uncoupled tuning of the double-well potential as shown in figure 3.2.

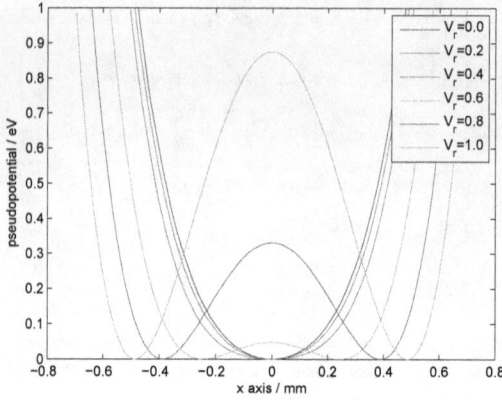

Figure 3.5: The calculated pseudopotential along the x-direction for an electrode structure similar to that shown in figure 3.4, as the voltage ratio $V_r = V_{adj}/V_{nom}$ is varied from zero to one. When V_r is zero, there is one well defined trap, as the V_r is increased, two traps appear. When $V_r = 1$ there are two well defined traps.

3.4.5 Electrode Geometry for Addressable Arrays

For the proposed architecture, shown in figure 3.6 is the conceptual progression going from a 2×2 static array of 2D traps to an addressable array. Starting from a normal array of point quadrupole ion traps (QIT) with a planar electrode geometry, each RF ring is segmented. These rings are then merged together, so that neighboring traps share a segmented addressable RF electrode. Arrays of such point ion traps with an adjustable RF could be designed in the rectangular pattern as shown, or in hexagonal or other patterns (see references [35, 162]).

The geometry of both point QITs [160] and arrays of such traps [163] has been extensively studied to optimize various parameters. Following on this research, the addressing RF electrode's length can be increased, so that the trap-depth efficiency factor κ_d (eq. 2.36) is improved. Increasing the addressable electrode's length requires the addition of a fixed RF electrode inside the 2×2 array.

This 2 × 2 addressable array could, in-principle, be scaled-up arbitrarily large. For the square lattice shown here, the number of adjustable RF electrodes scales as $2N$, where N is the number of trapping sites. Each trapping site needs minimally three individual electrodes to minimize the micromotion due to stray fields. This is can be done by biasing the RF adjustable electrodes with a DC voltage as well as adjusting the DC voltage of the circular trapping site electrodes. So that the necessary number of connections to the array scales as $3N$.

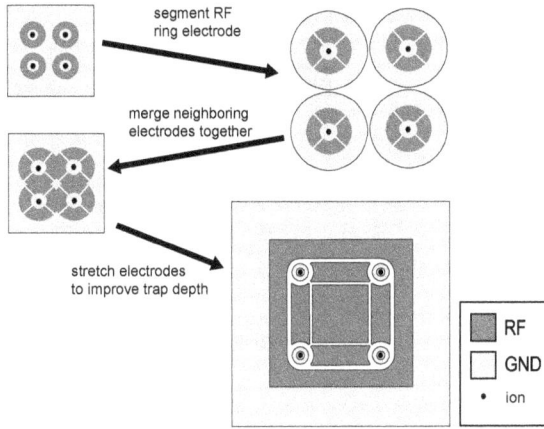

Figure 3.6: A representation of the conceptual process in going from a static 2-dimensional array of ion traps to an array where the neighboring ion-ion interaction can be addressed via the shared addressing RF electrode. Ions are trapped over each circular ground electrode when all RF electrodes are driven with equal voltage.

3.5 RF Displacement within a Single Point Trap

One of the promises of using adjustable RF electrodes with a 2D array of point ion traps is that the RF null positions of the individual ion traps can be changed. Specifically, if there is an adjustable RF electrode between two point ion traps in a 2D array (see figure 3.6, then the ions can be brought closer to each other by reducing the RF voltage on the adjustable electrode. This adjustment of the ion's position is called RF displacement.

In ion traps prior to this work, generally a single radio-frequency source was used to drive all RF electrodes. However, some experiments have applied differing RF voltages to each RF electrode so as to adjust the height of the ion above a surface electrode point trap [164–166] as well as adjusting the nodal line of RF in linear quadrupole ion traps [167]. Researchers have also used adjustable RF to convert two neighboring planar electrode linear traps into a single linear trap [168]. In 2D arrays of point ion traps, segmenting the RF electrodes and individually adjusting the RF voltage amplitudes creates additional degrees of freedom by which nearest neighbor interactions can be addressed and adjusted.

As described above in section 3.4.4, an adjustable RF electrode in between two surface electrode point traps allows a double-well pseudopotential to be tuned. One can model this potential with a multipole expansion of the electrode structure forming the two neighbors. Using such an approach qualitatively gives the same results as a quartic potential with an adjustable quadratic term. However, the mapping from the RF voltages to the quadratic and quartic terms is coupled and somewhat complicated (see eq. 3.41). In this section, the RF displacement of a single ion in its own trap is analyzed. This a valid approach, if the two neighboring traps are never merged. This approach greatly simplifies the analysis of RF displacement and the associated frequency tuning using an adjustable RF electrode placed in between two point quadrupole ion traps.

Below the quantitative effects of adjustable RF electrodes are considered, and how they can be used to perform RF displacement. First, the effect of a single RF voltage on the electric field in a trap is considered (3.5.1). Assuming that the phase on all RF electrodes is the same, RF displacement due to varying the RF voltage in a simplified ion trap geometry is then described (3.5.2). RF displacement in surface electrode traps is then considered (3.5.4). And finally the effects of a phase mismatch on the RF electrodes upon the average position and the micromotion of the ion is described (3.5.5).

3.5.1 Electric Field Due to a Single Electrode

If we are to explore varying the voltage amplitude of just one of the RF electrodes, it helps to first consider what effect the voltage on a single RF electrode has on the electric field in a very simple trap configuration. Figure 3.7a shows a trap in a two-dimensional space with four circular electrodes, where the geometric center of the trap is 0.5 mm away from the edge of the circular electrodes. The left electrode has $3/2$ V of RF voltage applied to it and other three electrodes have $-1/2$ V. So that the relative voltage on the left electrode is 2 V. Because no field lines go to the far-field, only the relative voltages on the electrodes affect the electric field. This electrode configuration could describe the cross-section of a 2D linear quadrupole ion trap.

Decomposition of the electric field and the voltages on the electrodes allows an analytical expression for the RF null position of the trap to be derived as a function of the voltages on the electrodes. The decomposition is shown in figure 3.7. A single voltage (fig. 3.7a) on the left electrode is decomposed into a dipole-like (antisymmetric) term (fig. 3.7b) and a quadrupole-like (symmetric) term (fig. 3.7c). The electric field arising from the dipole-like term, is approximately constant near the center of the trap. The electric field arising from the quadrupole-like term is zero in the geometric center of the trap, and is roughly proportional to the distance from the geometric center of the trap.

By simulating the dipole-like voltage configuration and the quadrupole-like voltage configuration, the electric field of these two configurations can be found. This then allows one to write the electric field E_x in the x direction as a function of the voltage V on left electrode and the x-position in the trap as,

$$E_x = V(a_x(x) - xb_x(x)/2), \tag{3.42}$$

where $a(x)$ is the zero-th order electric-field coefficient due to the dipole-like configuration and $b_x(x)$ is the first order coefficient (positively defined) of the quadrupole-like configuration. The coefficients are normalized to a 1 V difference on the electrodes. The gradient of the time-varying electric field $-\partial E_x/\partial x = \alpha_t$ (see sec. 2.1.1) is then related to the first-order coefficient b_x by

$$\alpha_t = V\frac{b_x(x)}{2}. \tag{3.43}$$

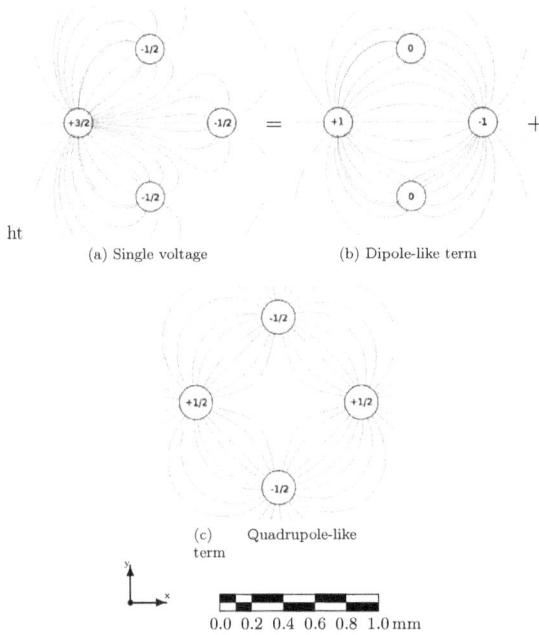

(a) Single voltage   (b) Dipole-like term

(c)   Quadrupole-like term

0.0 0.2 0.4 0.6 0.8 1.0 mm

Figure 3.7: Above is the decomposition of the voltage on a single electrode (a) into a dipole-like term (b) and a quadrupole-like term (c).

Using finite element simulations of this simplified trap geometry, the coefficients have been computed and are shown below in figure 3.8. The coefficient $a_x(x)$ changes from roughly 0.87 to $1.06\,\text{mm}^{-1}$ over the range -0.2 to $+0.2\,\text{mm}$. The quadrupole term ranges from about 5.2 to $5.35\,\text{mm}^{-2}$ over the same range. So that if 10% error in the electric field is allowed, a_x and b_x can be considered constant when the ion's position is $-0.2 < x < 0.2\,\text{mm}$.

3.5.2 RF Displacement in a Simplified Geometry

The location of the RF null (or trap location) can be computed by considering the RF voltage on the left (V_1) and RF voltage on the right (V_2) electrodes in terms of the coefficients a_x and b_x. Taking a_x and b_x as constants for now, the effect of RF displacement due to varying the voltage amplitude of one electrode can be estimated as follows. The x-component of the electric field due to a voltage V_1 applied on the left electrode is

$$E_1 x = V_1(a_x - b_x x/2). \tag{3.44}$$

Similarly, if a voltage V_2 is applied to the second electrode (on the right), the electric field generated is

$$E_2 x = -V_2(a_x + b_x x/2)). \tag{3.45}$$

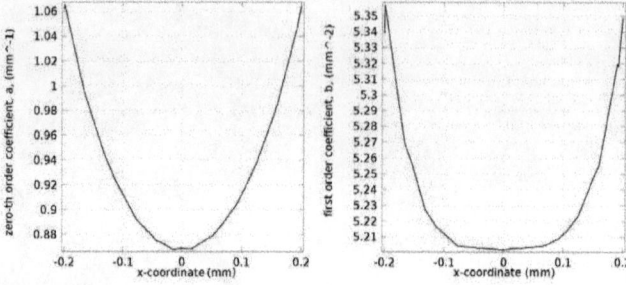

Figure 3.8: Results of a finite element method (FEM) simulation of the electric field inside a trap with geometry described by figure 3.7 given in terms of the zeroth order electric-field coefficient a_x (a) and first order electric-field coefficients b_x (b). Because (b) is the derivative of the electric field, which is a derivative of the simulated potential, the granularity of the FEM meshing results in some jaggedness of the coefficients a_x and b_x.

The field in the x-direction E_2, due to voltage V_2 uses the symmetry of the situation:

- The zeroth and first order coefficients have equal magnitudes for both electrodes.

- The voltage on the right creates the opposite dipole-like field, but the same quadrupole-like field.

If $V_1 = V_2$, then the trap is operating like a normal linear quadrupole ion trap (QIT) and the trap location is at $x = 0$. However, if the voltages are not equal in magnitude, but have the same phase, the trap location is found by setting the total electric field in the x direction $E_x = E_1 + E_2$ to zero and solving for the trap location x_t,

$$x_t = 2\frac{a_x}{b_x}\frac{V_1 - V_2}{V_1 + V_2}. \tag{3.46}$$

This gives the trap location as a function of the RF voltage amplitudes on the RF electrodes. It is completely generalizable, but care must be taken to consider that the dipole and quadrupole coefficients a_x and b_x cannot always be treated as constants. For the four-electrode trap given above, the variation in coefficients is about 10% over the range $\pm 0.2\,\text{mm}$. Consequently, one might expect the trap location to vary by 10% from that predicted by equation 3.46 when the ion is displaced 200 μm from the center of the trap.

3.5.3 RF Motional Frequency Tuning

The relationship of motional frequency of the ion ω_x, as the ion is subject to RF displacement in the x-direction can be calculated by considering the spatial derivative of the pseudopotential. The pseudopotential is proportional to the total electric field squared. To arrive at the motional frequency, first consider the total electric field in the x-direction formed by both the voltages V_1

and V_2 around the RF null point (x_t),

$$E_x = E_1 x + E_2 x = V_1 \left(a_x - \frac{b_x x}{2} \right) - V_2 \left(a_x + \frac{b_x x}{2} \right)$$

$$E_x(x = x' + x_t) = -\frac{b_x x'}{2} (V_1 + V_2). \tag{3.47}$$

where x' is the displacement of the ion from the RF null trap location x_t. For the 2D quadrupole ion trap described, the electric field in the y-direction at the RF null position is zero. If the motional axes are not tilted from the spatial axes, then the gradient of the electric field's y-component in the x-direction is also zero ($\partial E_y / \partial x = 0$), so that the frequency in the x-direction ω_x is then given by equations 2.23 and 3.43 as,

$$\omega_x = \frac{q|b_x|}{\sqrt{8} m \Omega_{rf}} (V_1 + V_2). \tag{3.48}$$

3.5.4 RF Displacement in Surface Traps

The example above, of RF displacement in a two-dimensional quadrupole ion trap, extends readily to more complicated two and three dimensional geometries, such as planar electrode point ion traps and even 2D arrays of such traps. It is necessary to calculate the coefficients a_x and b_x for the geometry in question. If the coefficients do not vary considerably over the range of motion desired, equation 3.46 can be used to estimate the RF displacement as a function of the voltage amplitudes on the electrodes and equation 3.48 can be used to estimate the motional frequency ω_x.

As a simple example, a 2D geometry consisting of a 3×1 array of surface-electrode ion traps could be analyzed in the same way as above. Figure 3.9 shows the decomposition of the voltage on just one addressable RF electrode (3.9a) into a dipole-like field (3.9b) and a quadrupole-like field (3.9c). The trapping site electrodes are spaced about 1.7 mm from each other. The location of the quadrupole RF null is about 0.5 mm above the middle trap electrode (see fig. 3.9c). Although the geometry here is very far from the simple geometry used to describe the RF displacement above, the fields can be decomposed in the same way.

A more realistic application of this decomposition is done for a 2D array of point ion traps below in section 6.1.3. There the electric field due to the voltage on the rest of the array's electrodes is also taken into account. However, the main idea is the same as shown here. The electric field due to a single voltage on one of the adjustable electrodes can be decomposed into a dipole-like field and a quadrupole-like field. The coefficients $a_x(x)$ and $b_x(x)$ can then be approximated as linear over some RF displacement range, so that equation 3.46 can be used to compute the location of the RF null as the adjustable voltage amplitude is varied.

3.5.5 RF Phase Displacement and Micromotion

The above discussion assumes that the amplitude of the RF voltage is varied while the phase of all RF voltages is the same. If instead, both of the electrodes' RF voltages have the same

(a) Single electrode voltage

(b) Dipole-like term

(c) Quadrupole like term

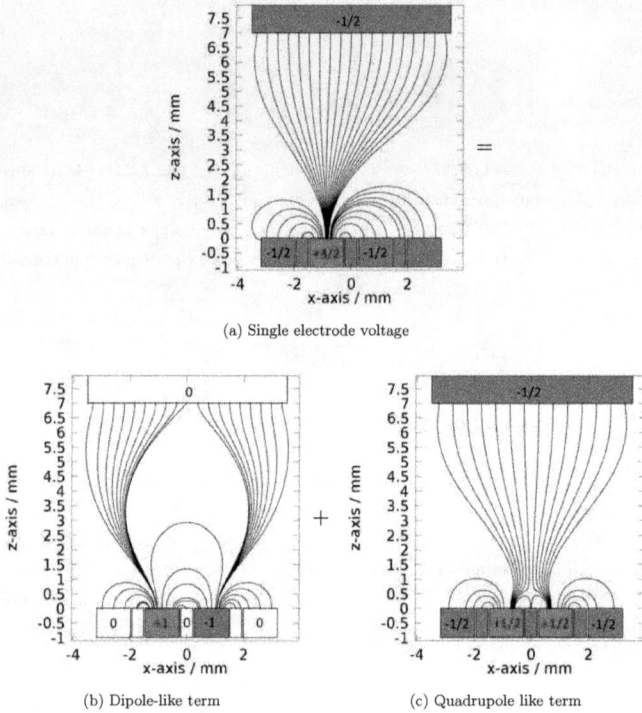

Figure 3.9: The decomposition of the field due to a single voltage in a 2D array of ion traps (a) into a dipole like field (b) and a quadrupole like field (c).

amplitude, but there is a phase mismatch between them, there will also be RF displacement but in addition there will be excess micromotion [136]. In this situation, there is no longer an RF null. Rather there is an RF minimum, at which, a trapped ion will experience excess micromotion due to this phase mismatch. In general, operating a trap with excess micromotion is considered a pathological condition to be avoided. However, it can be used to address the ions [137]. The ability to both increase the micromotion and perform RF displacement might then be desirable.

The position of the RF minimum, with the same voltage V on both RF electrodes, but with a phase mismatch ϕ, is given by,

$$x_{\min} = 2\frac{a_x}{b_x}\frac{1 - \cos(\phi)}{1 + \cos(\phi)}, \tag{3.49}$$

where it is assumed that one can approximate the coefficients a_x and b_x as constant. The excess micromotion experienced at this position, due to the phase mismatch is then given by

$$x_{\phi\mathrm{mm}}(t) = -\frac{qV\sin(\Omega_{\mathrm{rf}}t)\sin(\phi)}{m_{\mathrm{I}}\Omega_{\mathrm{rf}}^2}(a_x + x_{\min}b_x/2). \tag{3.50}$$

Thus, in the small angle approximation ($\phi \ll 1$), the excess micromotion due to a phase mismatch is proportional to the dipole-like coefficient a_x, the applied voltage V, the phase mismatch ϕ, and square of the inverse of the trap drive frequency Ω_{rf}.

3.6 Design of 2D Arrays

In this section, the design goals and limitations of various methods to scale up ions traps QIP are considered, with an emphasis on 2D arrays of ion traps. While the number of approaches for scaling up ion trap QIP is limited only by the imagination, a number of serious approaches have been proposed. These approaches are compared to the proposed 2D array of addressable point ion traps using adjustable RF electrodes.

The rest of this section is organized as follows: First the competing physical effects limiting the performance of 2D arrays of point ion traps are detailed (3.6.1). Then some of the alternative approaches to scale up ion trap quantum information processing are discussed (3.6.2). Methods to control and address the interaction in a 2D array of point ion traps are then laid out (3.6.3). Finally, a discussion of the various ways to address the individual ions and their nearest neighbor interactions is given (3.6.4).

3.6.1 Competing Physical Effects of 2D Arrays

Attempts at making ion trap arrays where the dipole-dipole interaction between ions is strong enough, while at the same time keeping the ions cold and maintaining a deep trap are often stymied by the physics of 2D trap arrays. Idenified here, are five characteristics, which need to be optimized in a 2D array of ion traps intended for quantum experiments and they are shown in table 3.1.

The first requirement of a deep trap can be done by building the trap electrodes close enough to the ions, with a suitably high RF voltage. The second item to consider is that of trap stability. As the trap is miniaturized, the trapping parameters q and a must stay within a stable range. The third item, that of a low heating rate, requires that the electrodes of the ion trap are not too close. The heating rate can be reduced by operating the trap in a cryogenic environment [169, 170]. Although the results of heating rate suppression in a cryostat are impressive, the predicted [132] and measured [171] heating rates in traps as small as tens of micrometers could still be too high for an experiment with an array of ion traps. For this reason, it is very desirable to find a way to trap the ions close to each other but still far from the electrodes, even if a cryostat is at hand. The last requirements for useful quantum experiments are that a strong dipole-dipole interaction, which is tunable and addressable is provided. Each of these items is discussed below and how they trade off with each other.

1.	deep trapping potential
2.	stable trap operation and low micromotion
3.	low heating rate
4.	strong dipole-dipole interaction for short two-qubit gates
5.	tunable and addressable interactions

Table 3.1: Performance criteria for 2D arrays of ion traps.

3.6.1.1 Trap Depth Scaling

The trap should have a depth capable of trapping ions from a hot oven (a typical atomic source) and holding them long enough to perform useful experiments. Typical experiments are performed in ultra high vacuum, though collisions do occur with the background gas. In addition, the heating rate of the ions, which rises as the trap is miniaturized, reduces the trapping time when cooling lasers are turned off. Experiments with micro ion traps (ion electrode separation $\sim 100\,\mu$m) [172] suggest that at typical pressures (10^{-9} to 10^{-10} mbar), single ion traps with a minimum depth of tens of meV can perform experiments.

The trap depth in the absence of DC fields for a general QIT trap can be written using equation 2.35 as

$$\phi_{\mathrm{D}} = \frac{\kappa_{\mathrm{d}} q^2 V^2}{4m\Omega_{\mathrm{rf}}^2 d^2} \tag{3.51}$$

where trap-depth efficiency factor κ_{d} is defined (see eq. 2.36) as the ratio of the trap depth of a given trap to the trapping depth of a perfect hyperbolic-electrode Paul trap, V is the trap drive's RF voltage, m is the ion's mass, Ω_{rf} is the trap drive's frequency, and d is the ion-electrode distance. For a planar-electrode point quadrupole ion trap (QIT) with a far-field ground, this efficiency is at most 2 percent [160], though this can be improved by a factor of 3-4 by simply placing a ground plane a small distance above the array [173]. It might seem that one could increase the trap depth simply by increasing the RF voltage, but the voltage is technically limited (especially in microtraps) to avoid electrical current between the electrodes.

3.6.1.2 Stable Trap Operation and Low Micromotion

In order for a quadrupole ion trap (QIT) to stably hold ions, it must be operated within the so-called trap stability regime (see sec. 2.1.2). To operate near the center of the stability regime ($q_x = 0.5, a_x = 0$), the ratio of the secular frequency ω_{x} to the trap drive frequency Ω_{rf} should be approximately 0.17. Smaller ratios are permissible with attention to DC fields in the trap, but larger values can easily become unstable [174]. For a point QIT without DC fields, this requirement can be written as (see sec. 2.1.2),

$$\frac{\omega_j}{\Omega_{\mathrm{rf}}} = \frac{q_x}{\sqrt{8}} \sim 0.17, \tag{3.52}$$

where ω_j is the secular frequency of the ion in the trap in the j-th direction and q_x is the trap stability parameter.

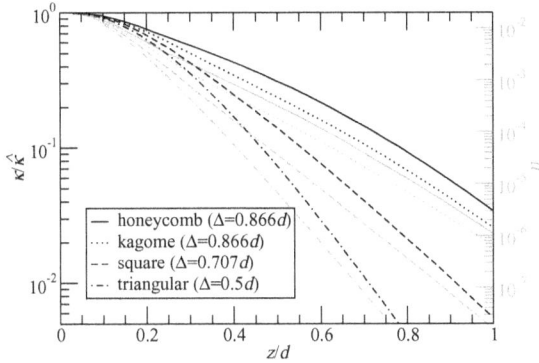

Figure 3.10: Normalized trap curvature and depth for various geometries of 2D trap arrays as given by Schmied [163]. The bottom axis shows the ratio of the ion height z to the inter-trap spacing d. The curves show the performance of various topologies of 2D lattices of ion traps, where the performance is given by the trapping depth efficiency η and the ratio $\kappa/\hat{\kappa}$ of the electric field's curvature κ to the curvature of an optimized surface ring-trap $\hat{\kappa}$ [160]. The efficiencies and curvatures drop-off exponentially with increasing z/d ratio. Reprinted figure with permission from R. Schmied, Phys. Rev. Lett. 102, 233002 (2009). Copyright 2009 by the American Physical Society.

3.6.1.3 Heating Rate Scaling

Heating of the ions can limit the performance of quantum-logic gate operations between ions, especially in microtraps [171]. And it can cause ion loss when the cooling lasers are turned off. It is therefore important to consider the scaling of the heating rate compared to the gate-operation time as the distance between the ions in a 2D array is reduced. Since the heating rate goes up as the ion-electrode distance is reduced (see appendix D), it might seen prudent to design the trap array so that the ions are far from the electrodes, but close to each other. Unfortunately, if this is done, the trap performance suffers drastically.

Figure 3.10 shows the normalized trap curvatures and depth efficiencies for various geometric aspect ratios of 2D trap arrays as given by Schmied et al. [163]. The bottom axis shows the ratio of the ion height z (our notation d) to the inter-ion spacing d (our notation a). The curves show the performance of various topologies of 2D lattices of ion traps, where the performance is given by the normalized trapping depth efficiency η and the dimensionless pseudopotential's curvature κ. If, in-order to minimize heating, the array of ion traps were to be designed so that the ions were close to each other, but far from the trap electrodes (z/d is large), the trapping potential would fall off exponentially (see fig. 3.10). This would make such a trap impractical to load and operate. This is in contrast to ions in segmented linear traps, where their segmented DC electrodes can allow multiple potential wells to be separated by distances smaller than the trap feature sizes [142, 175], since the ions are displaced along the line of RF null. This restricts the geometry of 2D trap arrays so that as the inter-ion spacing a is reduced, the ion-electrode distance d is necessarily also reduced.

When the constraints of trap stability, a minimum physical trap depth, a maximum applied RF voltage, and fixed trap shape are taken into account, it becomes necessary to increase the trap frequencies ω_x and Ω_{rf} linearly as the inter ion distance a, and consequently, the trap geometry is miniaturized ($\omega_x \propto 1/d$). This can be seen by examining the constraint equations (3.52) and (3.51). From equation (3.51), a fixed ϕ_D and κ_d, requires that $V \propto \Omega_{rf}d$. Substituting this into equation (3.52), it is then required that $\omega_x \propto 1/d$. As explained above, for an array constructed with well designed $\kappa_d > 1\%$ planar-electrode, point quadrupole ion traps, the inter-ion distance a is related to the ion-electrode spacing d ($a \geq 2d$) [163], so that the trap frequency is then proportional to the inverse of the inter-trap spacing; i.e. $\omega_x \propto 1/a$. This allows one to determine the scaling of the time required for a two-qubit gate, as the trap array is miniaturized.

The time for a two-qubit gate is given by equation 3.24 as,

$$T_{gate} = \frac{4\pi^2\epsilon_0 m\omega_x a^3}{q^2}. \tag{3.53}$$

When the relation $\omega_x \propto 1/a$ is inserted into this equation, the time for an entangling gate then scales as the square of the inter-ion distance a^2 instead of a^3.

For a static 2D array, the inter-ion distance a and ion-electrode distance d are proportional, it follows that, as the array is miniaturized and the ion is brought closer to the electrodes, the heating rate goes up. With array miniaturization, the gate time decreases as d^2 (since $a \propto d$). This poses a question regarding heating: If the gate-time is reduced as the trap is miniaturized as d^2, does the heating rate increase faster or slower than this?

The scaling of heating rates depends on the dominant physical mechanism [125] and can be expressed as a power law $\propto d^{-\beta}$ [132]. Strictly heuristically, and taking all the room temperature ion trap experiments as a whole (see appendix D), the distance scaling exponent $\beta \sim 2$. This allows one to argue that overall, miniaturization has not significantly deteriorated the coherent coupling of ions (normalized to the heating rate). Whether further miniaturization will improve the performance of quantum gates versus the heating rate cannot be known without identifying the dominant heating-rate mechanism in ion traps, which is still an open question. Conservative design then dictates that the trap array should be made small enough, but no smaller, so that an entangling gate can be performed, subject to other sources (non-heating) of error. This is also practical, as miniaturization is technically demanding.

The degree of miniaturization is then dependent upon the speed of two-qubit gates in trapped ion systems. For laser sources which perform gate operations on $^{40}Ca^+$ ions, high fidelity two qubit gates take approximately $50\,\mu s$ [176], though gate operations, which take as long as several hundred μs, have been performed [177]. 2D arrays of ions that are miniaturized enough so that a two-qubit gate operation takes on the order of $\sim$50-500 μs would allow for quantum experiments to be performed.

3.6.1.4 Gate Times between Traps with a Trapping Potential of 1 eV

Because the voltage applied to the electrodes is limited in magnitude by the breakdown voltage of surface traps (several hundred volts), the relations of equations (3.52), (3.53) and

(3.51) can be used to calculate the minimum time for an entangling gate as a function of ion-ion spacing in an array of microtraps (for a given κ_d and ϕ_D). By including the constraints of a practical planar ion trap ($\phi_D \sim 1\,$eV and $\kappa_d \sim 5\%$), the gate times shown in the third column of table 3.2 are significantly longer than those initially suggested by Cirac and Zoller [108], and greatly exceed times for high fidelity gates with ^{40}Ca$^+$ ions using known techniques. In the last column, the trap potential has been relaxed to $\sim 10\,$meV, resulting in gate times which could conceivably be used for entangling gate operations. In this relaxed trap, the trap frequency ω_x is much less than $1/7^{th}$ of the drive frequency Ω_{rf}, but trap stability can still be maintained with careful attention to applied DC biases. Unfortunately, a 2D array of ion traps with a $10\,$meV trap depth will be of only limited use because of a short experimentation time limited by ion loss. Frequent reloading of the trap is made all the more difficult by the large number of trapping sites. Because of these limitations, there are a number of significant issues in creating an experiment with a conventional 2D array of planar ion traps.

a (μm)	$\frac{\omega_x}{2\pi}$ (MHz)	T_{gate}(ms) $\phi_D \sim 1\,$eV	reduced T_{gate}(ms) $\omega_x{\prime} = \omega_x/10$, $\phi_D \sim 10\,$meV
1500	0.5	2200	220
375	2	140	14
100	7.5	10	1
50	15	2.5	0.25
25	30	0.6	0.06

Table 3.2: Example gate times for an entangling gate between ions in a 2-dimensional array for various inter-ion interaction distances a (first column). The trapping frequency ω_x is shown in the second column for a trap depth of $\sim 1\,$eV. In the third column are the corresponding gate times. In the last column, the trap has been relaxed by lowering the trap drive voltage so that the trap depth ϕ_D is $\sim 10\,$meV, while the reduced trap frequency $\omega_x{\prime}$ is $\omega_x/10$.

2D arrays of ion traps with a variable RF electrode placed between the trapping sites would, to some extent, get around these limitations. The overall trapping potential of the array can be made quite high, while the pseudopotential between trapping sites can be addressably lowered, and the ions brought closer, so as to increase the dipole-dipole interaction and reduce the time of a two-qubit gate. But this is not the only proposal to scale-up trapped-ion quantum information experiments. The next section discusses some of these alternative proposals.

3.6.2 Various Approaches to Scale up Ion Trap QIP

To better explain why the proposed variable-amplitude RF voltage electrode helps in the goals of 2D ion trap arrays, it helps to consider what are the key features which make them desirable for quantum physics experiments. For useful, large-scale quantum simulation and computing, thousands or even millions of qubits would be needed, which cannot be done in a traditional linear quadrupole ion trap. One possible path to scale up a linear ion trap is by the use of segmented DC electrodes [143]. The ions would then be shuttled around a segmented microscopic linear ion trap between various zones. Increasingly complex electrode structures, capable of shuttling ions between traps, have been produced with the aim of developing a fully functional and scalable quantum computer or quantum simulator [144, 178–183]. This approach is technically very demanding and it would be interesting if there was a simpler way.

An alternative proposal for scaling up quantum systems with ion traps was made by Cirac and Zoller [108] and uses a closely-spaced 2D array of ion traps. A highly detuned standing wave, which can create a state dependent force [108], or a Mølmer-Sørensen-type controlled-phase-gate [105] could be used to create a two-qubit gate between ions in neighboring traps. In both cases the dipole-dipole force between the ions would be the interaction allowing for an optical state-dependent force to generate the entanglement [142, 156, 175]. Entangled states could be created between neighboring ion traps, possibly allowing for large scale measurement based quantum computing [184] and simulations of quantum systems [28].

Efforts to use Penning traps [185–187] or optical dipole forces [188] to create a periodic 2D array of such microscopic ion traps are also underway. Two-dimensional arrays of 3D quadrupole ion traps [173] have been tested and proposals have been made to improve the trapping dynamics of such arrays [163] and increase the variety of physical systems which could be simulated [189]. Trap arrays could be fabricated with a planar geometry [179, 181], where each unit in the array would be a *point* trap [165]. Planar structures are desirable in that they allow the use of photolithographic techniques for fabrication. This permits miniaturization of the traps and scaling-up of the array through replication. An important feature of these arrays of point traps is that each trap uses the time varying (typically radio frequency) electric field to confine the ion in all 3 dimensions.

There are however several shortcomings with the above proposals for arrays of quadrupole ion traps, which have made them difficult to realize. One reason is the trade-off between the separation of ions in the neighboring trapping sites and the separation between the ions and the electrodes. For appreciable ion-ion coupling, the proposal of Cirac and Zoller [108] calls for bringing ions within tens of micrometers (or less) of each other. For most ion-trap array geometries this requires that the ions are also tens of micrometers from the electrodes. Unfortunately, if the ions are this close to the electrode materials, it could be that motional heating of the ions becomes very high [132, 190] compared to the ion-ion coupling. More practically, such small structures are difficult to build and connect to electrical sources. At these ion-electrode separations, accessing the ions with laser beams without grazing the surfaces also becomes challenging.

Ions are ideally kept at the RF null of a point quadrupole ion trap (QIT) so that they avoid being driven by the trap's RF electric fields. The point-like nature of the RF null in the 2D arrays of point QITs means that ions cannot be significantly displaced from the RF null positions without experiencing excess micromotion. This means that using DC fields to address and tune the coupling in 2D arrays of ion traps requires that careful attention is made to micromotion compensation. In general this approach would allow the trap frequencies to be adjusted, but not the ion locations. This limits the range which the dipole-dipole coupling can be adjusted (see below).

Because of the limitations of static (or DC controlled) 2D arrays of ion traps, it seems that keeping the ions far away from electrode materials, so as to avoid motional decoherence [132, 190], but close enough for an entangling gate [108], while maintaining an operable trap depth [163] does not appear particularly feasible. 2D arrays of planar-electrode quadrupole ion traps with dynamically-reconfigurable RF voltages have been suggested [191, 192] in which ions are displaced around to control their interactions. Others have changed the RF parameters in dust traps [193] or ion traps [167, 194] to control the position of the trapping site. These proposals allow for ions to be displaced or shuttled in two dimensions, and kept on the RF null, similar to the method proposed here.

The dynamically reconfigurable arrays proposed by Chiaverini and Lybarger [191, 192] would switch a large array of RF electrode pixels ($\sim$25 adjustable electrodes/ion) to allow for a complete reconfiguration of a planar trap providing for the ions to be shuttled around the chip. What is proposed here is more basic ($\sim$2 adjustable electrodes/ion) and allows for the addressing and adjustment of the interaction between nearest neighbors, where the overall topology of the trap array is not changed.

3.6.3   RF Addressing in a 2D Array

Here we quantitatively compare the proposed method to increase ion-ion interactions using variable RF voltages on the addressing electrode versus just reducing the voltage on the whole array. From equation 2.34, reducing the trap frequency, can be done by simply lowering the RF potential of the whole array. Another method suggested already, is to use a local modification to the pseudopotential in the region of two neighboring ions. If only the *pair* of neighboring ions of interest could be *addressed* with a lower RF drive amplitude, so that *their* interaction was increased, it might be possible to do experiments with an array of 2D ion traps. As explained in section 3.5, the lowering of an adjustable RF electrode placed between sites in the 2D array, allows for the ion's motional frequency to be lowered, while simultaneously bringing the ions nearer to each other. Both of these effects increase the dipole-dipole coupling relative to the motion of a single unperturbed trapped ion as given by equation 3.24. In the lowest order approximation, lowering the RF drive amplitude will cause the RF null position of a point trap, to be displaced proportional to the voltage amplitude (see eq. 3.46) and its frequency to be adjusted as per equation 3.48. The lowest order approximation of the trap frequency for the whole array is that it is proportional to the trap voltage 2.23.

A comparison of nearest-neighbor addressing via an adjustable RF electrode (as described in sec. 3.4) versus addressing the nearest neighbors with only shared frequency tuning is given in figure 3.11. The top figure shows both the frequency of coherent coupling of motion between the ion in each trap (dashed lines) and the normalized coupling vs. the heating rate (dotted lines), as a function of the trap frequency during addressing. The heating rate is assumed to be inversely proportional to the trap frequency ω_x squared, as it would be from a source of electric-field noise (see eq. 2.92) with a $1/\omega_x$ profile. One can see that, a trap where the nearest neighbors are addressed with RF displacement (upper dashed line) allows the coupling to be increased almost three orders of magnitude more than just with shared frequency tuning (lower dashed line). Furthermore, the RF displacement allows the normalized coupling rate vs. heating rate to be increased during the RF displacement (upper dotted line) vs. the shared frequency addressing (lower dotted line). The shared frequency addressing results in that the normalized coupling rate can only be decreased if the absolute coupling rate is increased. The lower figure shows the distance between the ions as the trap is addressed. Below about 1 MHz, the Coulomb repulsion pushes the ions away from each other, even as the distance between the RF nulls a_t is further

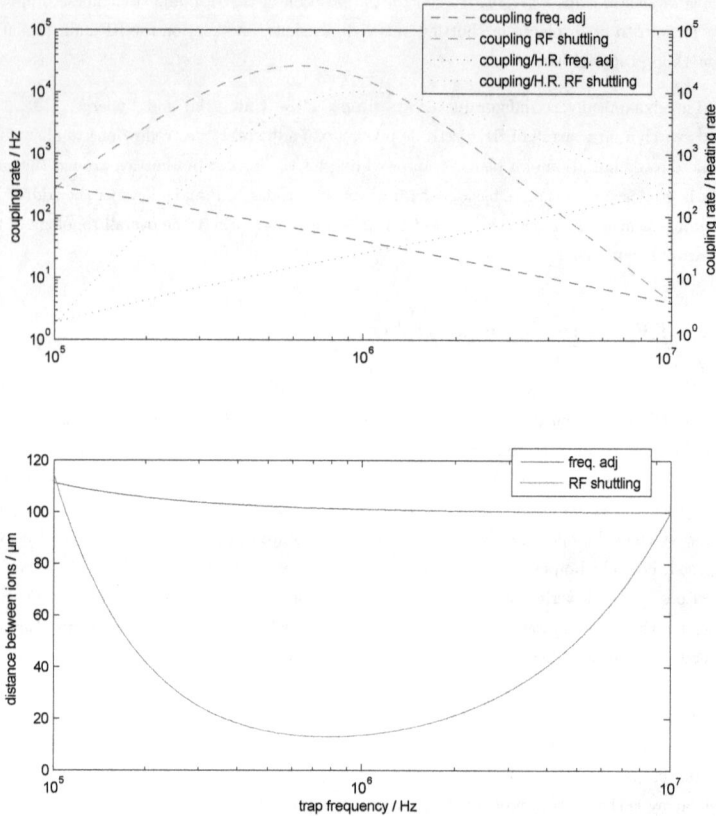

Figure 3.11: A comparison of addressing the interaction of two nearest neighbors in a 2D array, by either RF displacement or shared frequency tuning. The top figure shows both the coherent coupling rate of motion between the ion in each trap (dashed lines) and the normalized coupling rate vs. the heating rate (dotted lines), as a function of the trap frequency during addressing.. The heating rate is assumed to be inversely proportional to the trap frequency ω_x squared, as it would be from a source of electric-field noise (see eq. 2.92) with a $1/f$ profile. One can see that, a trap with, where the nearest neighbors are addressed with RF displacement (upper dashed line) allows the coupling to be increased almost three orders of magnitude more than just with shared frequency tuning (lower dashed line). Furthermore, the RF displacement allows the normalized coupling rate vs. heating rate to be increased during the RF displacement (upper dotted line) vs. the shared frequency addressing (lower dotted line).

reduced. The RF displacement model used here assumes that the double-well pseudopotential is never merged into a single well potential, so that the quartic term never comes into play (see section 3.5). Merging of the two traps into one, would have open up the possibility to increase the coupling even further.

3.6.4 Individual Addressing of Ions in the 2D Array

Addressing and controlling the coupling between ions in a 2D array is one of the main themes of this thesis. This is not the only way to address ions in an ion trap and it does not answer the need for addressing of the individual ions for single qubit operations. In principle, individual ions could be addressed using tightly focussed laser beams. These could be parallel to the surface [165] or at an angle to it. To avoid scattered light from the trap surface, particularly in view of laser-induced charging issues [175, 195], it is generally required to have holes through the substrate with either fibre [194] or free-space [196] delivery of light. One could also use transparent traps [197]. Alternatively, the ions could be addressed by altering specific transition frequencies. DC or RF voltages can be used to change the ions' motional frequency, or magnetic fields can be used to change the frequencies of electronic transitions [198, 199].

A simple method to address ions in a 2D array would be to use a laser beam which shines on the whole array, parallel to the surface and then use electronics to address the ion(s) motional frequencies. This can be done by adjusting the height of the ions above the trap or by tuning the motional frequency of the ions in question, so that the laser frequency is only resonant with the motional sideband of the addressed ion(s). Using static voltages to address a single ion in the array could be done simply by adjusting the DC voltage of the circular trapping-site electrode. However, when the ion is displaced from the pseudopotential minimum in a point quadrupole ion trap (QIT), it experiences an excess micromotion. The static voltage on the circular trapping-site electrode would induce considerable vertical micromotion of the addressed ion as well as horizontal micromotion on its neighbors, but if the aim is only to address a single ion, with a laser beam from the side, this may not be a concern.

3.7 Towards a Running 2D Algorithm

To test the 2D array, a relatively simple experiment, which takes advantage of the proposed RF addressable 2D array architecture (see sec. 3.4), could be performed. The simplest incarnation of a 2D array would be a 2×2 arrangement of ion traps. For instance, the measurement based computing described above in section 2.12.5 requires that 2D array of qubits is prepared in a cluster state. A cluster state in a 2×2 array could be produced using a simulated nearest neighbor Ising interaction [200]. Measurements could then be made in the various basis to implement a CNOT gate or other simple quantum circuits. The standard techniques of trap loading, Doppler cooling, electronic state preparation, sideband cooling, single qubit gates, and measurement (all described in chapter 2) could be used. The pairwise 2D entanglement (sec. 3.2.6) and gate operations (sec. 3.2.8) would allow for a complete set of logic operations.

Performing even such a simple experiment would require exact control of the ion's displacement to perform the necessary interactions. The RF addressing method has the potential to avoid heating the ions during displacement as the pseudopotential would be altered adiabatically by the raising and lowering of the adjustable RF. However, stray fields could cause the trap frequencies to drift. The technical details then become quite important as to whether an architecture, based upon 2D arrays of RF addressable interactions, would be able to perform as hoped. A microscopic addressable array, where the inter-ion spacing a is 100μm (reducible to 50μm), might be able to perform such an algorithm.

In order to test the ideas of adjustable RF and addressing the ions in such an array of traps, several preliminary experiments were carried out. The next chapter describes the general experimental setup for trapping ions. After that the experiments to test the ideas of 2D arrays are described in the next three chapters in order of increasing technological difficulty.

Chapter 4

General Experimental Setup

In this chapter, the technical details for trapping $^{40}\text{Ca}^+$ ions in a vacuum chamber, laser cooling and imaging them are described. This setup will then be assumed in the following chapters (unless otherwise specified) to characterized a series of ion traps. This is possible because, for quantum optics experiments with trapped $^{40}\text{Ca}^+$ ions, certain facilities can be used for several types of experiment. For instance, the cooling and fluorescence lasers at a frequency of 397 and 866 nm depend on the atomic level structure, so that they can be used on multiple experiments (with beam splitting cubes) or reused for experiments with different trap designs. Similarly, a single vacuum system can be used to test a series of ion traps, so long as after opening the chamber and installing the next ion trap, the appropriate vacuum level is reached.

The laboratory, while often not considered as part of the experimental setup provides a stable environment for the experiments. For the experiments described herein, above all, the laboratory must be temperature stable, or the many optical elements will require constant tuning. For the experiments described here, the temperature was kept at a constant 20° C±1° C.

First described in section 4.1 is the vacuum setup used for testing 2D arrays of point quadrupole ion traps. Next the laser sources for photoionizing and controlling the states of $^{40}\text{Ca}^+$ are detailed in section 4.2. The details of a laser stabilization cavity are then given in section 4.3. The imaging and fluorescence detection of the ion is described in section 4.4. And lastly the experimental control is described in section 4.5.

4.1 Vacuum Setup

For experiments with planar-electrode ion traps, a UHV vacuum chamber was constructed. Figure 4.1 shows the vacuum configuration. The trap was mounted into a Kimball physics octagon[1], with coaxial feedthroughs[2] mounted to the top flange and a viewport on the bottom flange. The coaxial feedthroughs provided electrical connections to the individual electrodes and the viewport on the bottom allowed imaging via an objective and camera. Viewports on the side

---

[1] Kimball MCF800-SphOct-G2C8
[2] Accuglass™ 25D-5CX2-450

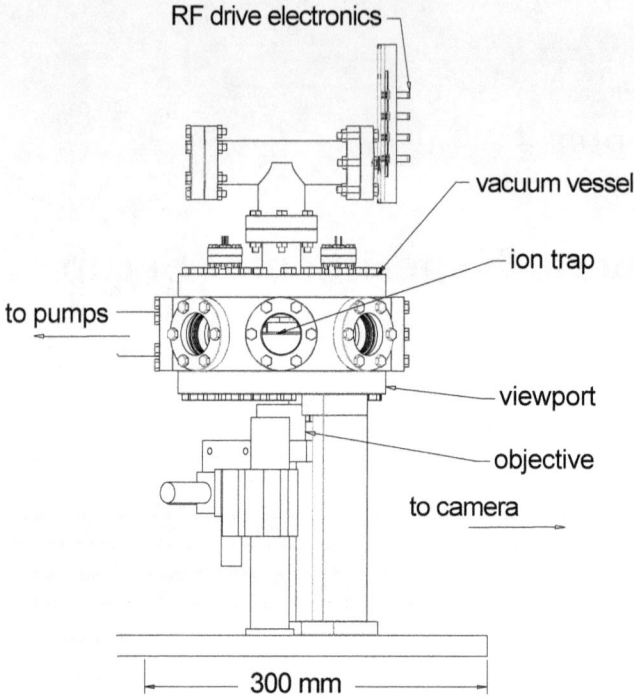

Figure 4.1: Above is a drawing of the vacuum setup for testing planar electrode ion traps. Lasers for cooling and photoionization have access through the viewports on the side of the vacuum vessel. Imaging of the ions is done through a viewport on the bottom of the vacuum vessel, through an objective. The vacuum pumps on the left include a Ti-sublimation pump and an ion getter pump. RF feedthroughs on the top allow for the trap to be driven by the RF electronics.

allowed for laser access in the plane of the trap array. Calcium atoms were provided by an oven[3] mounted inside a welded bellows port aligner[4] on the side of the octagon. A Ti-sublimation pump and an ion getter pump were used to maintain a pressure of approximately 10^{-10} mbar during the experiment. It is estimated that an ion at a pressure of 10^{-10} mbar will experience an inelastic Langevin collision rate [201] of approximately $\gamma_{\text{Langevin}} = 0.004 \, \text{sec}^{-1}$ and an elastic collision rate of about $0.03 \, \text{sec}^{-1}$ [202]. So that a minimum cooled ion lifetime of several minutes to hours (depending on the chemical reactivity of the background gas) is expected.

[3] Alvatec™ AS-3-Ca-150-C
[4] UHV Design PA-35H

transition	wavelength (nm)	laser type	locking	model
$4s4p\,{}^1P_1$-continuum	375	free running diode	none	bare diode
$4s^2\,{}^1S_0$-$4s4p\,{}^1P_1$	422	SHG ECDL	none	844 nm DL Pro
$S_{1/2}$-$P_{1/2}$	397	SHG ECDL	FP cavity	SHG Pro
$S_{1/2}$-$D_{5/2}$	729	ECDL	FP cavity	DL Pro
$D_{5/2}$-$P_{3/2}$	854	ECDL	FP cavity	DL 100
$D_{3/2}$-$P_{1/2}$	866	ECDL	FP cavity	DL 100

Table 4.1: Laser systems for photoionizing calcium and controlling its electronic levels. The first two laser systems allow for calcium I to be photoionized. The last four laser systems allow for all the electronic states in figure 2.5 to be controlled. All commercial laser sources were obtained from Toptica.

4.2 Laser Setup

Lasers and their associated systems to photoionize calcium I (see figure 2.9) and to control the states of calcium II (see figure 2.5) are described here. In total, six diode based laser systems to photoionize and control the state of ${}^{40}Ca^+$ ions were setup. Table 4.1 lists the different laser systems and their characteristics. The 375 and 422 nm lasers allowed for the photoionization of calcium via a resonant two-step process [80]. The last four laser systems allowed for all the states in figure 2.5 to be reached or pumped out.

For stable experiments, the 397, 729, 854, and 866 nm laser systems were all locked to a Fabry-Pérot (FP) stabilization cavity (see sec. 4.3) with a Pound-Drever-Hall (PDH) lock [203]. Each laser system had a double-pass acoustic optical modulator (AOM), so as to allow the frequency and power to be adjusted during the experiment. The AOMs were controlled by a personal computer, described below in section 4.5. The laser systems themselves are diagrammed in figures 4.3-4.5 and described in more detail next.

The 397 nm laser system in figure 4.3 gives an overview of the primary components necessary to control the polarization, frequency, and power of the laser for delivery of coherent light to ${}^{40}Ca^+$ ions. The 397 nm lasing source is a 794 nm external cavity diode laser (ECDL) with a tapered amplifier and second harmonic generation (SHG) cavity capable of supplying more than 60 mW of light. These components were available from Toptica Photonics as a complete system[5]. In order to lock this laser to the optical cavity, a small amount of the available light was sent through a blue EOM[6], used to create sidebands at 9.7 MHz, to the optical cavity. A small amount of laser light was also sent to a wavemeter[7], which had a multiplexer, so that up to 8 laser wavelengths could be measured. The lasers were stabilized with the help of electronic modules[8]. The frequency of the locked laser could be tuned by adjusting the length of the optical cavity via a piezo ring inside. The stabilized 397 nm light was then sent to the various experiments in the laboratory. At each experiment, a double pass AOM[9] setup allowed the power and the frequency of the laser to be adjusted. Because this AOM was polarization sensitive, it was necessary to

---

[5] Toptica Photonics AG TA-SHG Pro
[6] Photonics Technologies EOM-01-10-U
[7] High Finesse WS7
[8] Toptica PDD and FALC
[9] Brimrose QZF-80-20-397

use a prism and a pick-off mirror to complete the double-pass of the AOM and send the light to the experiment. The control electronics for the AOM driver, are described in appendix B.

The 854 and 866 nm lasers both use a similar setup as the 397 nm laser, but with a simplified optical path shown in figure 4.4. The sidebands for the PDH lock were generated directly by modulating the power to the laser source, which was an external cavity laser diode[10]. Each laser source had a power output of up to 50 mW. A small amount of the available power was then sent to the wave meter and optical cavity to stabilize the wavelength and measure it. At the experiment, a double pass AOM[11] was setup, using a polarizing beam splitter and a $\lambda/4$ phase plate to overlap the beams.

The 729 nm laser was setup similarly as the 397 nm laser and is shown in figure 4.5. An external cavity laser diode with a tapered amplifier[12] provided up to 500 mW available output power. A small amount of available light was sent through an EOM[13] to the optical cavity for stabilization. The Pound-Drever-Hall detector for this laser was built using a digital function generator[14], a splitter[15], RF amplifier[16] and RF mixer[17]. A double-pass AOM[18] was setup similar to the 854 and 866 nm laser paths to vary the frequency and power to the trapped ion.

Using the optical cavity, described in section 4.3, it was possible to lock the laser with a 3 kHz linewidth and a drift of about 80 Hz/sec, as measured with a beat between another 729 nm laser locked to a ^{40}Ca$^+$ S-D transition. In figure 4.2 is a plot of the beat between the two lasers as a function of time. To improve the system's stability further, a second vertical high-finesse cavity [204, 205] was added to the system, allowing for the laser to be brought down to linewidths of about 1 Hz and a drift of approximately 5 Hz/sec (see Erhard [206]).

The lasers beams access the ion by entering through the viewports on the side of the vacuum vessel. Because the vacuum vessel was elevated above the optical table, the optics for focussing and steering the laser beams were mounted on breadboards approximately 200 mm above the table. All laser beams were overlapped into a single optical beam with the use of wavelength selective filters as follows. A Semrock SEM-LD01-375/6-25 filter allowed for the 375 nm laser to be overlapped with the 422 nm, these two lasers could then be overlapped with the 397 nm light with a Semrock SEM-FF01-395/11-25 filter. These three beams could then be overlapped with the combined 854 and 866 nm light via a long pass SEM-BLP01-594R-25 filter. The 854 nm and 866 nm beams were overlapped via a polarizing beam splitter (PBS) so that the two beam's polarization were orthogonal to each other.

---

[10] Toptica Photonics DL 100
[11] Crystal Technology, Inc. Model 3200-124
[12] Toptica Photonics Syst TA DS w/DL Pro
[13] Qioptiq PM25
[14] SRS DS345
[15] Minicircuits ZMSC-2-1+
[16] Minicircuits ZFL-1000+
[17] Minicircuits ZAD-1
[18] Brimrose TEF-270-100-.729

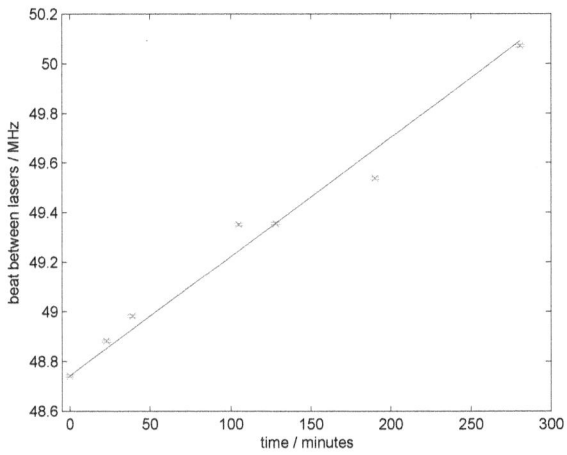

Figure 4.2: Drift rate of 729 nm laser locked to the optical cavity. The laser drifts at a rate of about 80 Hz/sec when locked to the cavity.

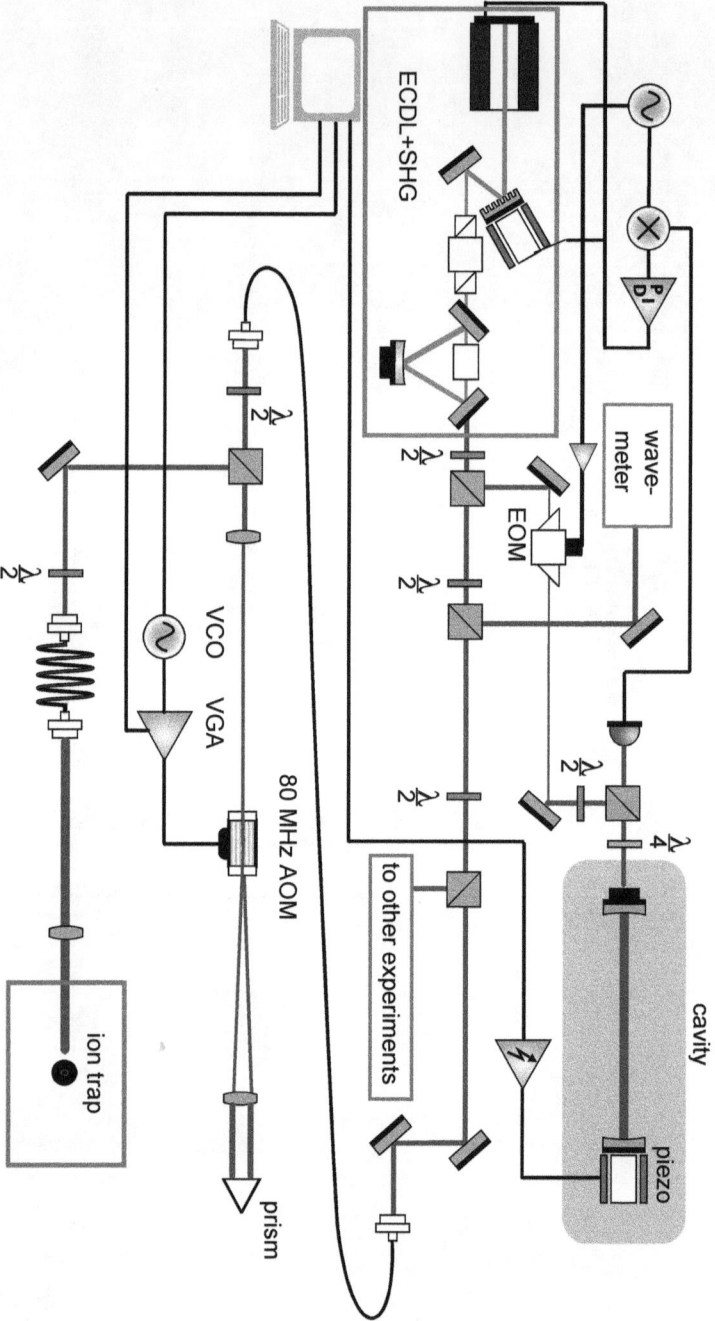

Figure 4.3: Schematic of 397 nm laser setup. It is used for Doppler cooling and fluorescence based state detection.

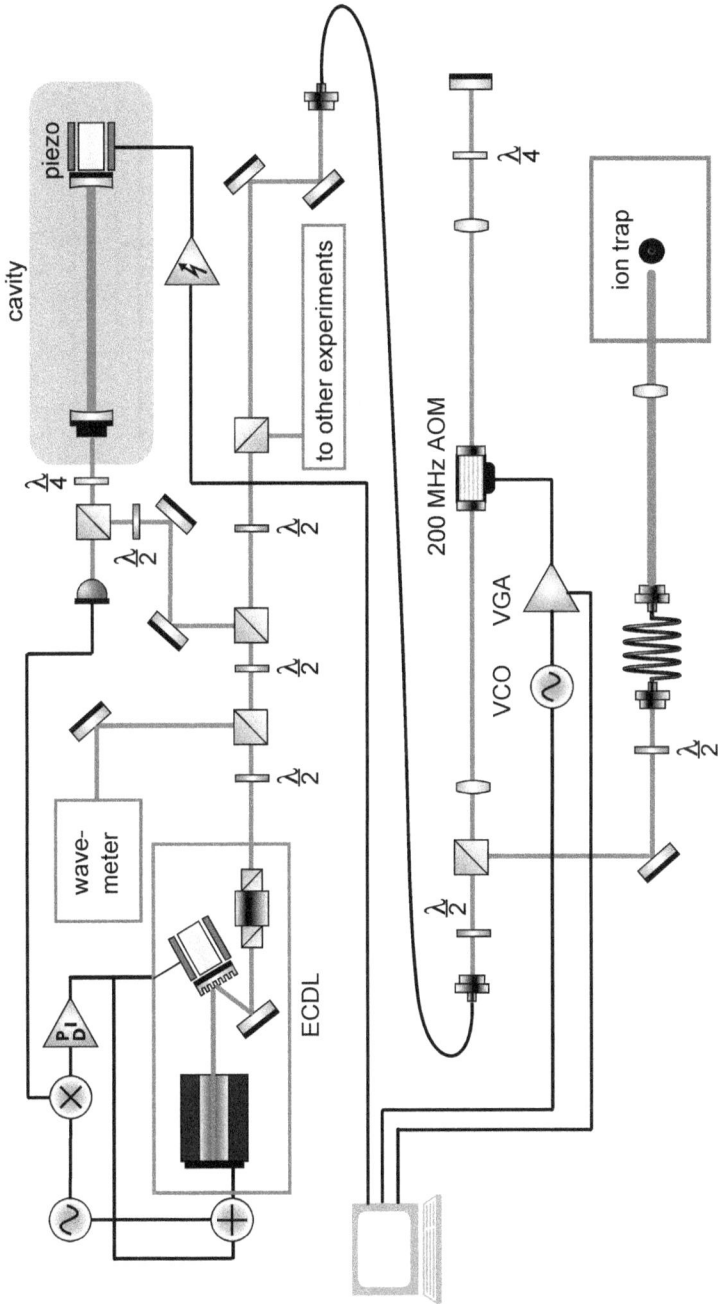

Figure 4.4: Schematic of 866 nm and 854 nm laser setups. The 866 nm light is used to pump the ion out of the $D_{3/2}$ state during Doppler cooling and the 854 nm light is used to pump the $D_{5/2}$ state out to the $P_{3/2}$ state. Details in the text.

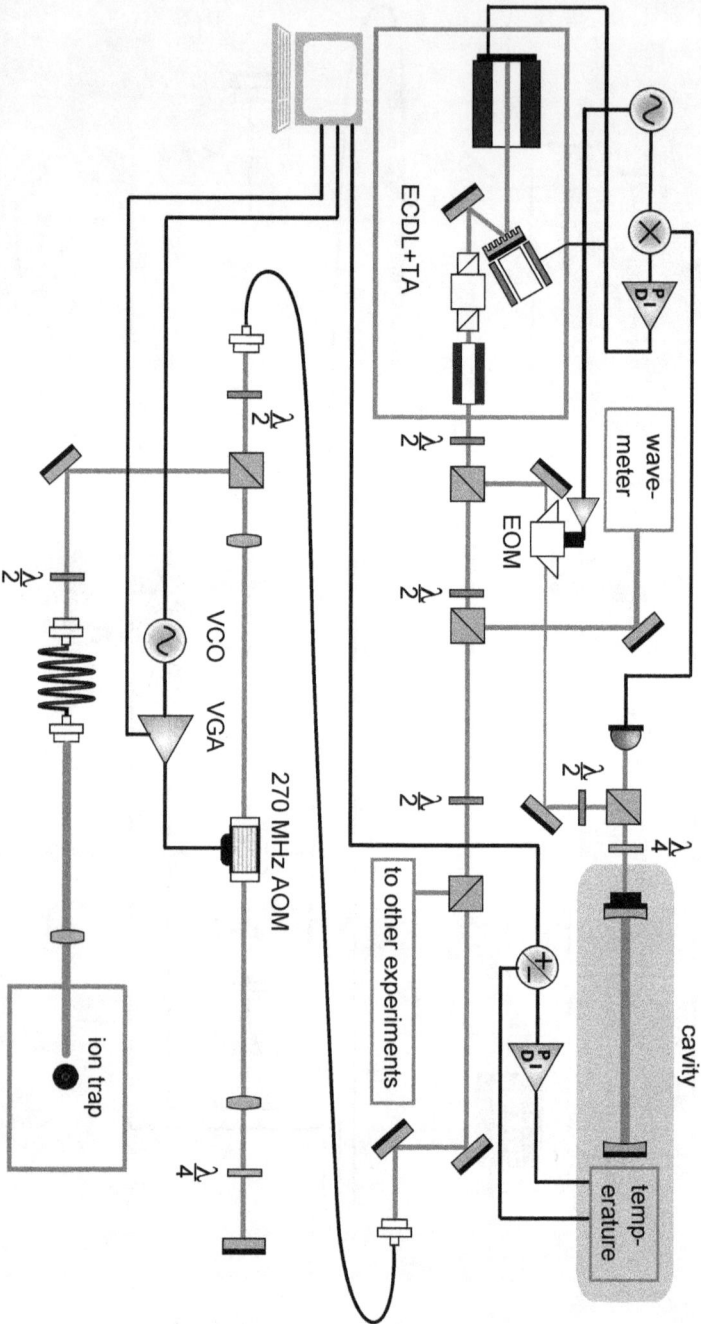

Figure 4.5: Schematic of laser setup for the 729 nm transition. The 729 nm light is used to drive transitions between the $S_{1/2}$ and $D_{5/2}$ states. Details in text.

4.3 Optical Cavity

The "optical cavity" is actually 4 Fabry-Pérot (FP) interferometers built on the same block of Zerodur™ low temperature coefficient expansion glass built into a vacuum system. The housing is designed to shield the cavities from vibration and temperature fluctuations so that when the lasers are locked to them via a Pound-Drever-Hall lock [203], the lasers have a stable, but tunable, frequency. For the three dipole transition lasers, 397, 854 and 866 nm, it was possible to vary the length of these FP interferometers because one of the mirrors was mounted to a piezo ring, to which a voltage could be applied. The design of the cavity was based upon an existing Innsbruck design [207]. However, some practical details were added or changed as follows.

First the vacuum ion pump[19] is built within the temperature stabilized region of the optical cavity assembly. Second, the temperature control is done by Peltier elements with passive heat sinks, which can convect over 200 W of heat away. Dual digital Peltier temperature controllers[20] capable of stabilizing the temperature to approximately $2\,\mathrm{mK}/\sqrt{\mathrm{minute}}$ were used on each side of the cavity assembly. The whole assembly had several layers of insulation and thermal capacitance so that the settling time constant of the FP interferometers due to any temperature change occurring outside of the cavity was about 24 hours. The expected temperature stability was expected to be better then $1\,\mathrm{\mu K}/\sqrt{\mathrm{minute}}$ at the cavity due to random fluctuations at the temperature controller.

In figure 4.6 is a cross-section of the cavity assembly, with call-outs to some of the major components. The outer box is composed of black polyoxymethylene (POM/acetal) plastic (not labeled). Just inside this, is a 30 mm thick sheet of polystyrene for insulation (not labeled). The outer heat shield is fabricated from 20 mm thick aluminum plate (call-out 41) and temperature controlled by Peltier elements. Inside this heat shield is the vacuum system with view ports (call-outs 29-36), which mounts on the outer heat shield via POM clamping brackets (call-outs 37-40). Inside the vacuum system is an inner heat shield (call-outs 18-21,25), which is composed of 20 mm thick aluminum and sits inside the vacuum system on six 12.7 mm diameter Viton™ balls (call-outs 22-28). Inside the inner heat shield is the 100 mm Zerodur™ glass spacer (call-out 1) upon which the Fabry-Pérot interferometers are built. This spacer rests inside the inner heat shield on eight pieces of 8 mm thick Viton™ cord. Each interferometer is composed of a pair of mirrors, one flat, and one with a radius of 0.5 m[21]. The curved mirrors are glued to the piezo ring assemblies (call-outs 10-13) via a Macor™ ring with an ultra-violet curable epoxy[22]. The piezo rings were electrically contacted using a silver filled conductive epoxy[23]. The mirrors have a reflectivity of 99.5% for the dipole transitions and 99.95% for the 729 nm laser. Each window or viewport in the assembly is tilted so as to avoid unwanted etalons. The vacuum system is kept at a pressure of approximately 10^{-8} mbar by the vacuum pump (call out 36).

A test of the quality of the 729 nm laser locked to this cavity was done by overlapping the laser light with another remote 729 nm laser, which was locked to a calcium ion, on a fast photodiode. The beat note between the two lasers provided an upper bound on the drift rate

---

[19]Varian™ 2 L/s
[20]Ing.Büro R. Tschaggelar TEC
[21]Laser Components GmbH
[22]Dymax OP-67-LS
[23]Epo-Tek H20E

and linewidth of the 729 nm laser. When locked, the short term drift of the 729 nm laser was measured to be approximately 80 Hz/sec (see fig. 4.2) and the linewidth was measured to be no more than 3 kHz. To vary the frequency of the 729 nm laser, the temperature of the cavity could be changed, where the frequency tunability at room temperature was 57 MHz/° C.

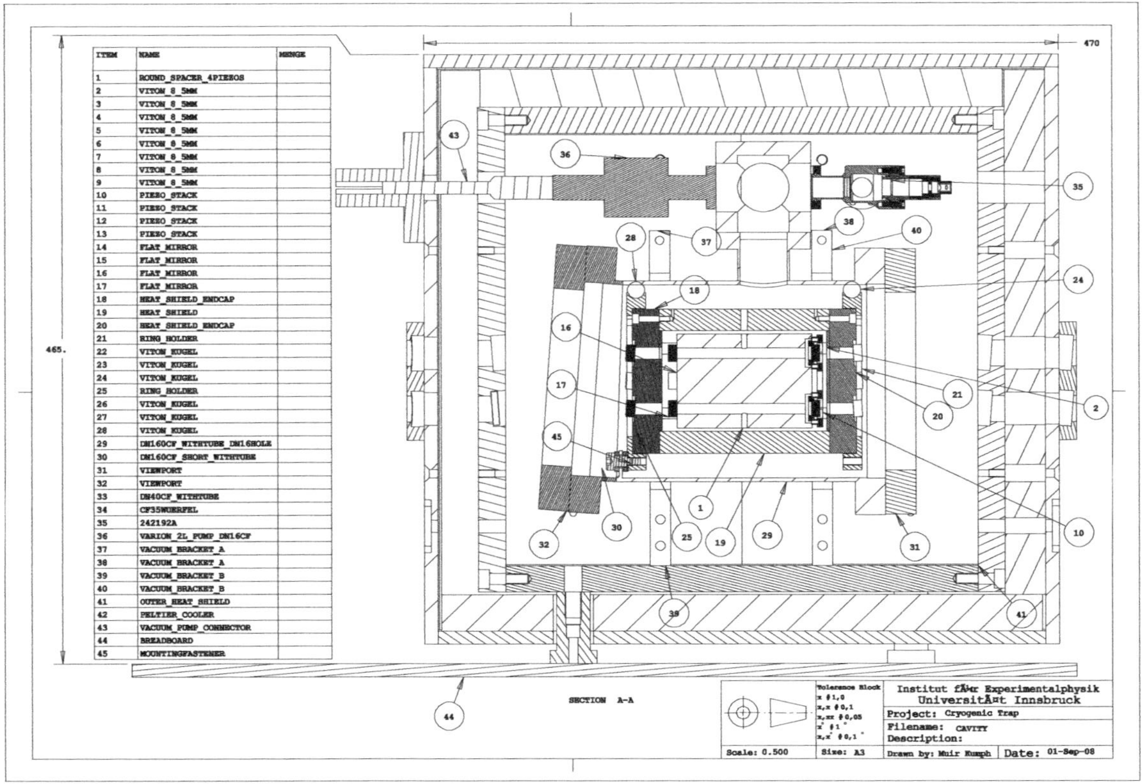

Figure 4.6: Layout of laser stabilization cavity. The cavity has 4 sets of mirrors, where each pair form a Fabry-Pérot interferometer which is used to stabilize a laser using the Pound-Drever-Hall lock method. More details in text.

4.4   Imaging and Florescence Detection

In order to image the $^{40}Ca^+$ ions, a charge-coupled-device (CCD) camera, made by Andor[24] and a custom objective[25] with a working distance of 70 mm were used. To gather time series information about the photons, the image path was split with a 50/50 beam splitter so that a photomultiplier tube (PMT)[26] could gather fluorescence photons.

The count the photon clicks generated by the PMT and register their time arrivals, either a timer/counter card[27] installed in the experiment controller computer was used, or a time-correlated single photon counter (TCSPC)[28] was used. For ion loading or state detection, the timer/counter card was sufficient, but for the reheating measurements the TCSPC was necessary so as to have a high-resolution record of the ion's fluorescence.

To minimize stray light to the PMT, three Semrock™ filters were used on both the camera and photomultiplier tube (PMT). Two stacked narrowband filters[29] allowed the calcium II fluorescent light to be transmitted. These could be removed, to expose a broadband filter[30], to allow detection of calcium I atom fluorescence.

4.5   Experiment Control

The control of the various experimental devices was done from two personal computers running the Microsoft Windows™ operating system. One was responsible for running the camera software. The other had digital to analog (D/A) output cards[31] which allowed voltages to be sent to the AOM sources (so as to control the power and frequency of the lasers), the DC trapping site voltages, or the DC bias of the RF electrode. A Labview program allowed the experimenter to control the analog voltages, and turn on and off the lasers via a graphical user interface. The RF electrodes were controlled by a set of DDS based function generators[32] phased locked together with a common 10 MHz source. The whole laboratory was supplied with a GPS referenced 10 MHz source[33], so that all precision frequency sources were supplied with a common clock.

---

[24] iXon3 885 S/N X-3719
[25] Sill Optik Model S6ASS2241
[26] Sens-Tech P25PC
[27] National Instruments PCI-6601
[28] Picoharp™ 300
[29] Semrock SEM-FF01-395/11-25
[30] SEM-FF01-417/60-25
[31] National Instruments PCI-6703
[32] TTi TG5001
[33] Jackson Labs Fury GPS Disciplined Frequency Standard

Chapter 5

Dust Trap Array "Dusty"

A 2×2 array of point quadrupole ion traps meant to trap and control the interactions between charged dust particles (called "Dusty") is described here. This allowed the computer simulations of the electric field for a similar ion trap to be verified qualitatively. It showed that varying the voltage amplitude of the time varying potential displaced the RF null, as described above in section 3.4.

The rest of this chapter is organized as follows. First the results of the computer simulations of the electric field and resulting pseudopotential, depending on the voltage amplitude of the addressable electrode, are given in section 5.1. Then the results of building and operating a dust trap with the same geometry are given in section 5.2. Finally a discussion of the results is given in section 5.3.

5.1 Simulations of a Small Addressable Array

The fabricated design of a small array is shown in figure 5.1. In between each trapping site is an addressing RF electrode with voltage V_{adj}, which serves to modify the local trapping potential. This shared electrode allows one to lower the trap frequency of the two neighboring traps. In the limit of the voltage amplitude going to ground, the two traps merge to become a single linear trap with an axial frequency of almost zero. Surrounding this 2×2 array is a fixed-voltage-amplitude RF ring electrode at V_{nom}, and surrounding that is a ground electrode. To improve κ_{d}, a ground plane is positioned above the array (not shown in figure 5.1).

To individually adjust (here called *address*) the interaction between local pairs of point ion traps in an array, an RF electrode between the point quadrupole ion traps (QIT) is inserted, and its voltage is adjusted. By attenuating the RF voltage on this *addressing* electrode, the electric field in between the two point QITs is reduced. In this way, a neighboring pair of ions in the 2D array can have their interaction increased, leaving the rest of the 2D array of ions isolated and with a deep trapping potential. It also permits a larger overall trap depth when compared to lowering the RF drive of an entire array. Additionally, it allows for the addressing of just pairs of nearest neighbors, by bringing just the addressed pair closer to each other and lowering their shared trap frequency.

For analysis, a 2D array of ion traps with an addressing RF electrode between the trapping sites was designed. This was then simulated using finite-element analysis to compute the electrical field. The pseudopotential of the trap was then computed with equation 2.25 to determine the trap depth ϕ_D. Because these traps are harmonic in the region close to the RF null, the trap frequency, and therefore the trap stability was estimated by fitting a parabolic function to the pseudopotential ϕ.

5.1.1   Simulations of Dusty for Ions

Shown in figure 5.2, is the pseudopotential as it varies in space for a 2×2 addressable array of ion traps containing $^{40}Ca^+$ ions with an intertrap spacing of $a=6$ mm. The simulations show that when the addressing RF electrode potential V_{adj} is equal to the fixed RF electrodes potential V_{nom}, the 2×2 array has 4 separate point quadrupole ion traps (figure 5.2i). As expected, by attenuating V_{adj} from 100% of V_{nom} to ground, the two point quadrupole ion traps bordering the addressing electrode are morphed into a linear trap. Because of the asymmetry of the trapping potential as V_{adj} is attenuated, the trapping sites move closer to each other (see figure 5.2ii). As V_{adj} falls, the two trapping sites move rapidly (with V_{adj}) toward each other until they are above the boundary of the addressing RF electrode and the ground electrode. As V_{adj} goes to zero, the two trapping sites do merge into a single trapping site, albeit with a very low shared motional frequency ω_x. The ion interaction could be significantly increased, even while maintaining the ions in well defined separate point quadrupole ion traps, by making the addressing electrode shorter than that shown here. The trap-depth efficiency factor κ_d would then necessarily be reduced. The simulations show that a ground plane a distance $a/2$ above the surface of the this planar trap array also improves κ_d from 1.7% to 6.7%.

In the simulations shown here, the addressing electrode length (plus the electrode gap) is about 90% of the distance between the home trapping sites, which means that the ions can remain in separate traps, while the distance between them is reduced by $\sim 10\%$. The gate time between point quadrupole ion traps (QIT) scales with the cube of the distance, so that even in this long aspect-ratio trap, the gate time could be about 30% less with addressable electrodes than without. This is in addition to the increased interaction caused by attenuating the RF voltage, which reduces the secular frequency ω_x in both the conventional and addressable arrays of point QITs. The time for this displacement, assuming it is done adiabatically (i.e. without heating the ions), would be approximately 10 times the period of the slowest secular frequency of interest. Since the secular frequency scales approximately linearly with V_{adj}, even the slowest traps in table 3.2 could be RF addressed adiabatically in just microseconds. For instance, a 2D array with 100 μm inter-trap spacing and 75 μm long addressing electrodes should be able to perform an entangling gate in 400 μs, with the adiabatic addressing of the adjustable RF electrode taking on the order of 10 μs. While the addressing electrodes add some technical complexity, the increased interaction between neighboring sites and the deeper absolute trapping depth of the whole array means experiments could be significantly easier to perform.

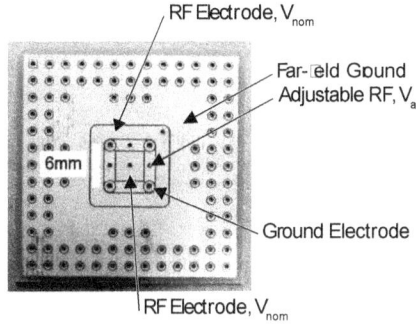

Figure 5.1: Geometry of the electrodes used for the simulation of trapping ^{40}Ca$^+$ ions in a 2×2 array of addressable ion traps as well as for the trapping of charged dust particles. Above each circular ground electrode is a point quadrupole ion trap. The trap is implemented on a circuit board and has the same physical layout as an industry-standard PGA101 integrated-circuit chip carrier so that it can be plugged into a standard socket.

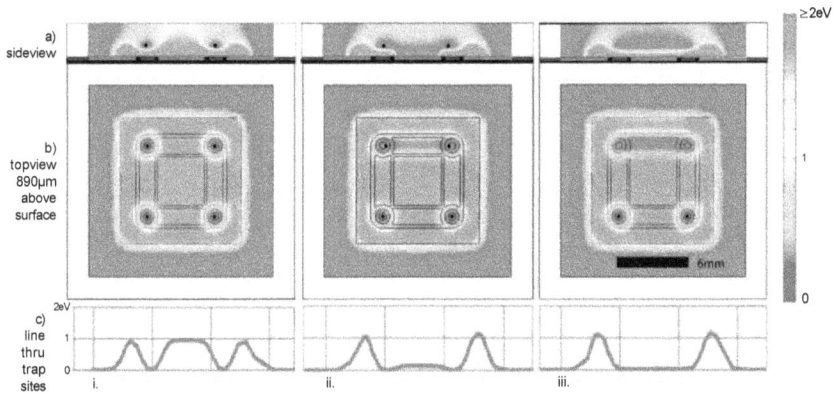

Figure 5.2: Series of simulations for trapping ^{40}Ca$^+$ ions in an addressable array of point ion traps with an inter-trap spacing of 6 mm. Top: a) series of simulations showing a slice through the pseudopotential field in eV through the middle of the two trapping sites which share an addressing electrode. Middle: b) series of simulations showing a slice though the pseudopotential 890 μm above the surface of the trap. Bottom: c) The pseudopotential along a line connecting the two trapping sites. From left to right (i-iii), the potential of the addressing electrode V_{adj} is ramped from 100% of V_{nom} (215 V 10 MHz) to 0 volts. Initially (i), the two ion traps are fully separated and form two well-isolated point quadrupole ion traps. In between (ii), V_{adj} is 43% of the nominal trap voltage and the trap frequency has been reduced. (iii) V_{adj} is at ground and the two addressed ion traps in the array are formed into a linear ion trap.

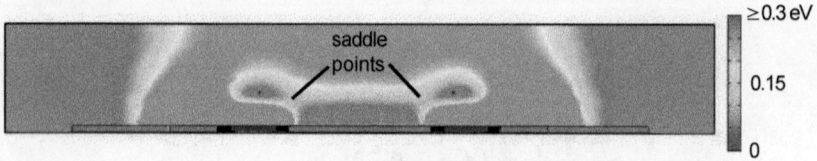

Figure 5.3: A closer look at the simulation results (from figure 5.2.a.ii.) when V_{adj} is at 43% of V_{nom}. Here the color scale has been set so that the saddle points are easier to see. A trap appears over the addressing electrode. The trapping locations are indicated by dark points in the spherical trapping potentials. The ion-ion distance has been also reduced by 10% compared to the distance with full voltage on the addressing electrode. The potential of the saddle point (here $\sim$0.1 eV) between the new trap over the addressing electrode and the point quadrupole ion traps falls with the square of the addressing electrode's voltage. However at all times the ions are trapped within the array by a strong potential.

Shown in figure 5.3, is the pseudopotential when V_{adj} is 43% of V_{nom}. A third trap opens up over the addressing RF electrode. The trapping locations are indicated by dark points in the spherical trapping potentials. The inter-ion distance has been also reduced by 10% compared to the distance with full voltage on the addressing electrode. The potential of the saddle point (here $\sim$0.1 eV) between the new trap over the addressing electrode and the point quadrupole ion traps falls with the square of the addressing electrode's voltage. However, unlike the trapping potential of a conventional array of point QITs with a weak trap depth, the ions in this array have a strong trapping potential in the overall array. The addressing electrode allows for the interaction between point QITs to be increased while avoiding the problem of a low trapping potential, which leads to ion loss. As V_{adj} is decreased to 0 volts, the third trap rises up and connects the two point QITs with nearly a line of RF null. In this way the two formerly point QITs are morphed into a linear QIT.

5.2 2×2 Dust-Trap Array

A printed circuit board[1] trap, shown in figure 5.1, was used to trap charged dust particles (Lycopodium spores) in air as a proof of principle. The trap was placed in a test fixture inside a plastic acrylic box to shield it from air currents (see fig. 5.4). The electrodes were driven by variable autotransformers, which were AC coupled to the trap to allow for bias DC voltages. A wire mesh was positioned 3 cm above the trap, with a voltage of 150 V applied to it, which provided gravity compensation for the charged dust particles. A green laser pointer was used to illuminate the setup.

5.2.1 Trapping and Morphing Traps with Charged Dust Particles

Shown in figure 5.5 are a series of photographs taken of clouds of dust in the charged particle trap Dusty. The voltage on the addressing RF electrode is lowered (pictures i-iii) and shows how

---

[1] copper on FR4 (frame retardant class 4) epoxy laminate

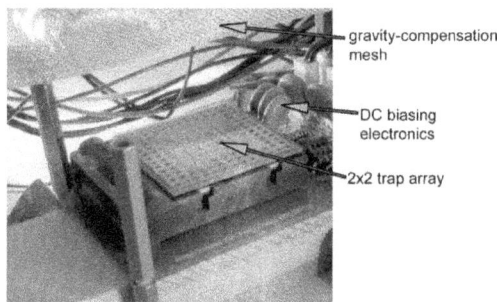

Figure 5.4: Photograph of the test fixture for trapping charged dust particles. The RF electrodes are driven at a voltage of 230 VAC at 50 Hz, with the amplitude tuned via autotransformers. High-pass filters allow for bias voltages to be applied to all RF electrodes. The entire assembly is housed in an acrylic box to shield it from air currents.

the two point-like traps are morphed into a single linear trap. Figure 5.5.i shows two point quadrupole ion traps, each with several dust particles. Figure 5.5.ii shows the cloud of particles in each separate trap have enlarged and moved toward each other as the two traps are relaxed. Finally, in figure 5.5.iii is a linear trap with a chain of dust particles.

The parameters used for trapping charged dust particles were different from those needed when trapping ions, because dust particles have a much lower charge to mass ratio. Instead of a 10 MHz RF drive, just 50 Hz is required. Despite the low frequency, this is still termed the RF drive in analogy with ion traps. The secular frequency of the charged particles could be seen and is ~ 8 Hz. Because of the small charge to mass ratio of the charged dust particles, they are affected by gravity and require a static field to pull them upwards, so that they are on the RF quadrupole null. This was accomplished by applying a compensation voltage to a mesh installed above the trap array.

5.2.2   Importance of Bias Voltages on Radio Frequency Electrodes

The charged dust particles can be steered or displaced by DC bias voltages on the RF drive electrodes. When two traps are in a linear trap configuration (figure 5.5.iii), a DC bias on an RF electrode can displace the charged dust particles from one area of the trap to another. During the process of morphing the two point QITs into a linear QIT, a small DC bias voltage (~ 2 V) was also used to keep the charged dust particles, which were vibrating with micromotion, from falling into the third trap (see fig. 5.3) which opens up over the addressing RF electrode. In addition, DC bias voltages can be used to minimize micromotion caused by the particles being pushed out of the RF quadrupole null by other fields, such as gravity or stray electric fields. Because of how useful the DC bias voltages on the electrodes were, it is desirable to also provide this capability in the electrode drive electronics when trapping ions.

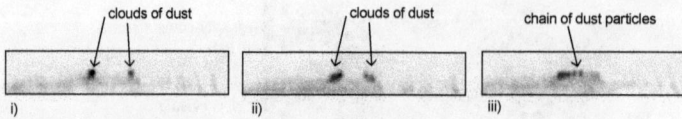

Figure 5.5: Photographs taken of the charged-particle dust trap. The voltage on the addressing RF electrode is lowered (pictures i-iii) and shows how the two point-like traps are morphed into a single linear trap. i) Two point quadrupole ion traps, each with several dust particles. ii) The cloud of particles in each separate trap have enlarged and moved toward each other as the two traps are relaxed. iii) linear trap with a chain of dust particles.

5.3   Discussion

The dust trap results showed that varying the amplitude of the 50 Hz drive for the electrodes allows the trap location to be tuned. When the voltage of this electrode was zero, the trap was morphed into a linear trap. The space charge of the dust particles caused the particles to form a linear chain. These results suggested that the same type of trap adjustment should be possible with ions.

Chapter 6

Ion Trap Array "Folsom"

In order to test the ideas of RF displacement and addressing of ions using an adjustable RF electrode, 4×4 trap array "Folsom" was built and tested. The inter-trap spacing of the array was 1.5 mm. This allowed the design to be made using a printed circuit board technology, but meant that loading multiple sites with a laser-sheet, or driving coherent dipole-dipole interactions between trapping sites was not to be done. Instead, single ions were loaded in sites of the array, and the ability to RF displace single ions in more than one dimension was demonstrated.

This chapter is laid out as follows. First the design of Folsom is given in section 6.1. Next the design and testing of the RF resonators for operating the adjustable RF electrodes is described in section 6.2. Next the setup of these RF resonators and associated electronics is described in section 6.3. The results of loading ^{40}Ca$^+$ ions in Folsom and adjusting the trapped ions with the adjustable RF are given in section 6.4. And finally a discussion of the experimental results is given in section 6.5.

6.1 Design of 2D Array Folsom

The methods of designing, predicting the performance of, and constructing the trap array Folsom are given here. First the design is reviewed and the basic simulations of the 4×4 array are shown (6.1.1). Then the method of fabrication is detailed (6.1.2). Finally, detailed simulations are given which show the expected RF displacement due to adjusting the RF voltage on nearby segmented RF electrodes (6.1.3).

6.1.1 Folsom Geometry and Pseudopotential Simulations

To trap ^{40}Ca$^+$ ions, a 4×4 square array of ion traps, Folsom, was designed so that the inner 2×2 array was fully adjustable and addressable. Shown in figure 6.1 is a photograph of the trap before being put into vacuum. The RF drive was segmented into 26 separate electrodes. This is was done in order to minimize micromotion both at the point trapping sites and the

linear traps possible between each pair of point quadrupole ion traps (QIT), as well as to create
an axial trapping frequency in the linear traps. Because of the large number of electrodes (see
figure 6.2) only the inner 2×2 part of the array was wired up to be fully adjustable, while the
outer 12 trapping sites provided a periodic boundary condition for the inner array. Folsom is very
similar in geometry to the 2×2 array described in chapter 5, with the exception of having more
trapping sites. The basic shape of the electrodes is the same, but there are more of them, and
the RF electrodes have been further segmented to allow for DC biases to provide micromotion
compensation, to displace ions and to impose an axial trapping frequency, when two point QITs
have been morphed into a linear QIT.

Figure 6.3 shows a simulation of the pseudopotential of the 4×4 array. It is qualitatively
similar to the 2×2 array. With a drive voltage of 125 V at 10 MHz, the ions are held within the
array with a trap depth of about 0.5 eV. A ground plane 1.5 mm above the surface improves κ_d
by about a factor of 4, and should provide shielding from stray charges in the setup. Figure 6.4
shows a close-up cross section of the trapping fields in the trapping array Folsom. The field lines
are shown in black and the pseudopotential is shown as color coded contours (same scale as in
figure 6.3). Because of the 3D nature of the electric-field simulation, the field lines disappear into
and out of the cross-section plane. The RF and DC ground electrodes are seen on the bottom
of the figure.

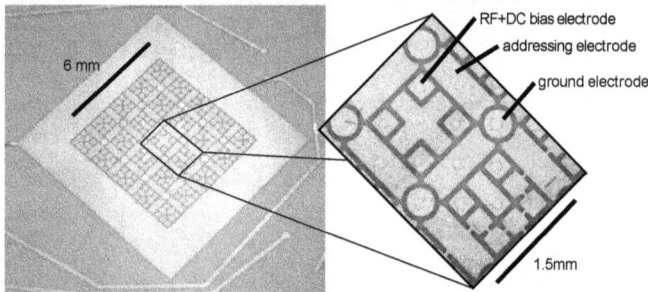

Figure 6.1: Fabricated trap. Inside the diamond area, is a 4×4 array of planar point quadrupole
ion traps. A microscope close-up of the middle 2×2 section of the trap shows details of the
electrode structure. The nominal trapping sites (circular electrodes ∅400 µm) are 1.5 mm from
each other.

6.1.2   Ion Trap Fabrication

Folsom was designed to determine whether tunable RF electrodes could be integrated successfully into an array of ion traps. The use of separate adjustable electrodes requires vias
integrated into the electrode structure. A printed circuit board technology was used for this
purpose [196]. This technology had a limitation of 50 µm traces and 50 µm gaps. To make the
trap as deep as possible for a given applied voltage, a ratio of 4 to 1 for the RF electrode's length
to the trapping-site electrode's radius was chosen [160].

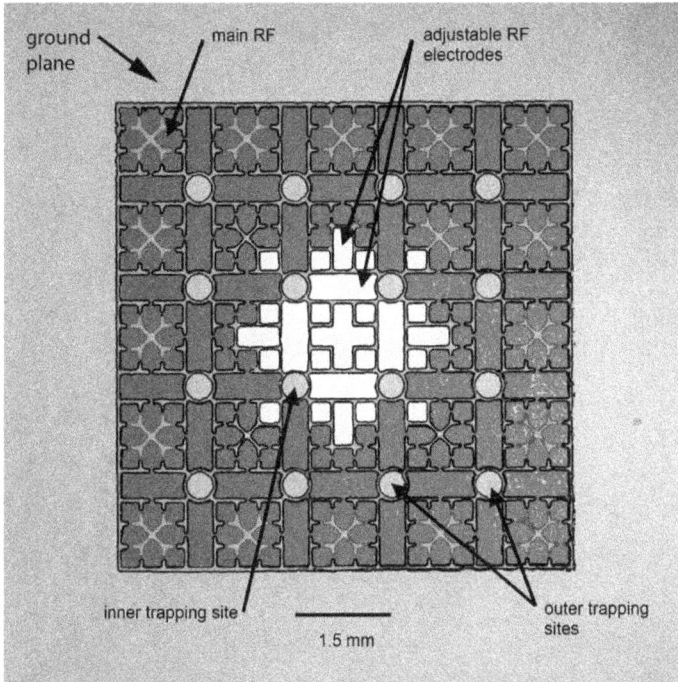

Figure 6.2: False color image of fabricated trap layout. The nominal trapping sites (circular electrodes $\varnothing 400\,\mu m$) are 1.5 mm from each other. The main RF is colored darker than the adjustable RF electrodes (bright). DC electrodes are colored light brown. The inner 2 × 2 trapping sites have many individually adjustable electrodes so that micromotion can be minimized.

Folsom's fabrication steps are shown in figure 6.5. First the electrode structure was etched into a printed circuit board (PCB) material [1] with copper cladding[2] (see Fig. 6.5a)[3]. The PCB was then plated with 10 μm gold using a sulfite gold solution[4]. A detail of the surface after gold plating is shown in figure 6.5b. The board was then mounted to a 1.6 mm thick substrate with pins (Mill-Max 3116), which were crimped into the vias to so as to support the PCB and connectorize it. The completed trap is shown in figure 6.5c. The trap was then placed into a specially made PCB filter board, which was then plugged into a cable break-out board as shown in figure 6.5d. These coaxial cables then were connected to the coaxial UHV feedthrough described above in the vacuum setup.

---

[1] Rogers™ 4350b 35 × 35mm 170 μm thick substrate
[2] 18 μm thick copper
[3] Andus Electronic Gmbh
[4] Transene TSG-250

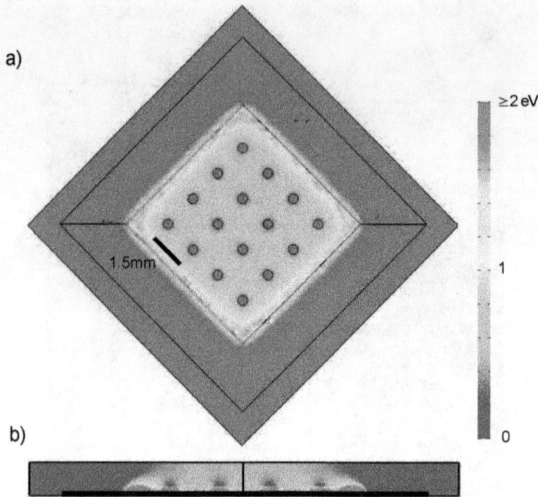

Figure 6.3: Simulation of Folsom trapping ^{40}Ca$^+$ ions showing the pseudopotential (eV) 375 µm above the surface of the trap (above) and also from the side diagonally through the trap. A ground plane sits 1.5 mm above the surface of the trap. The trapping depth is about 0.5 eV. All RF electrodes are fully driven with a voltage amplitude of 125 V and a frequency of 10 MHz.

6.1.3 RF Displacement Simulations

In order to extend the methods for quantifying the displacement due to adjustable RF electrodes developed in section 3.5 to traps with more than 2 RF electrodes, such as 2D arrays of addressable ion traps, then any extra RF fields from the other electrodes must be taken into account. Figure 6.6 shows the electrode structure of the Folsom array. The electrodes RF1, RF2, and G1 are the relevant electrodes for computing the dipole and quadrupole coefficients around the trapping site electrode G1. The residual RF field from the rest of the RF electrodes (colored red), must also be taken into account.

First the simulations are run, so that the dipole-like and quadrupole-like coefficients are found for the two adjustable RF electrodes (RF1 and RF2) on either side of a trapping site (electrode G1). Then the electric field due to voltages V_1, V_2 on the adjustable electrodes (RF1 and RF2) is simply

$$E_{\text{adj}} = V_1(a - bx/2) - V_2(a + bx/2). \tag{6.1}$$

If the trapping site is located in a periodic array such as Folsom, so that it contains periodic even symmetry of the RF electrodes, the effect of applying the same voltage to all the other RF electrodes will be to create a weak quadrupole field at the center of the trapping site. This extra field E_{EF} is related to the voltage on the rest of the array's RF electrodes V_{nom}, by

$$E_{\text{EF}} = -V_{\text{nom}}cx, \tag{6.2}$$

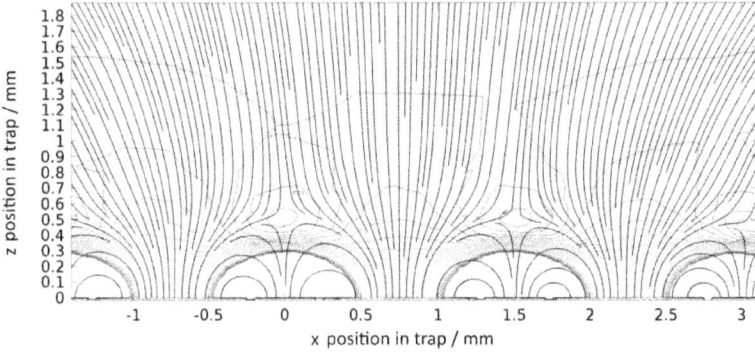

Figure 6.4: A cross section of the trapping fields in the trapping array Folsom. The field lines are shown in black and the pseudopotential is shown as color coded contours (same scale as in figure 6.3). Because of the 3D nature of the electric-field simulation, the field lines disappear into and out of the cross-section plane. The RF and DC ground electrodes are seen on the bottom of the figure.

where c is the positively defined quadrupole-like coefficient of the rest of the array's electrodes, and x is the distance from the center of the trapping site under examination. The total field is then the sum of the field from the adjustable RF electrode and the extra field from the rest of the RF electrodes in the array. By setting this total field to zero and solving for the trap location x_t, we arrive at

$$x_t = 2 \frac{a(V_1 - V_2)}{b(V_1 + V_2) + V_{\text{nom}}c}. \tag{6.3}$$

Figure 6.7 shows simulations of the electric fields as a function of the x-position around the center of a trapping site for the various applied voltage configurations. A procedure analogous to that given in section 3.5.2 is used, so that the simulations are performed as follows:

For the dipole-like electrode configuration 6.7a, corresponding to coefficient a, the two near RF electrodes have a voltage of +1 V and -1 V while the rest of the array is set at ground.

For the quadrupole-like configuration 6.7b, corresponding to coefficient b, the two nearby electrodes have a voltage of +1 V, while the nearby ground electrode and ground plane are set at 0 V. Because the coefficient b is defined for ±1 V, the gradient should be multiplied by 2 to equal b. A line is fitted to the computed electric field, so as to determine the gradient of the electric field (which is then the co-efficient b).

For the residual array RF 6.7c, term c, all the RF electrodes, except for those neighboring the trapping site are set to +1 V, while all other electrodes are set at 0 V. In order to compute the coefficient c, a line is fitted to the electric field so as to find the gradient. The extra field has the effect of producing a quadrupole like field which is only about 5% compared to the adjustable RF electrode's quadrupole term.

For a 2D array where all voltages are initially set to the same value, RF displacement can be accomplished by attenuating the power to the adjustable RF electrode. By setting the dipole

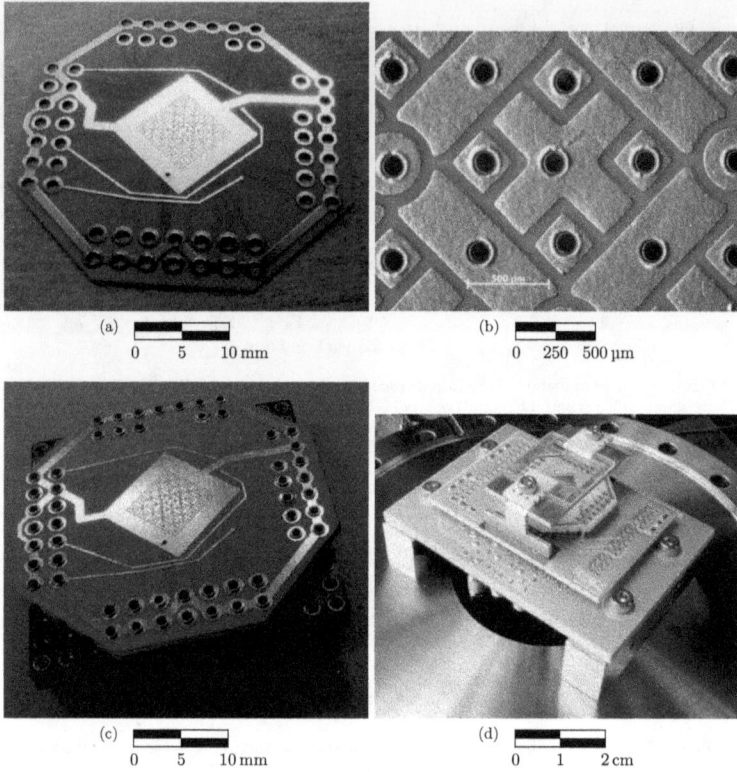

Figure 6.5: Fabrication of the Folsom ion trap. (a) printed circuit board. (b) detail of the center electrodes' structure after plating with gold. (c) connectorized trap. (d) trap plugged into the vacuum filter board and connector board. A transparent ground plane is mounted 7 mm from the trap surface. The entire assembly is then installed in the vacuum chamber with the trap facing downwards.

term a and the quadrupole like term b in equation 3.46 to the values from the simulation results in figure 6.7, the expected displacement of the RF null in mm is given by

$$x_t \simeq \frac{5(1 - V_r)}{15 + 14.3V_r},$$

(6.4)

where V_r equals the relative magnitude of the adjustable RF electrode's voltage (RF1), ranging from 0 to 1 and the displacement distance is measured in mm. For instance if the power to the electrode is attenuated by 4 dB, so that the V_r is about 63% of the maximum voltage. Then one should observe an RF displacement of about 77µm from the initial trapping position. Comparing this number with the simulation results in figure 6.7 shows that parameters are constant in this region to within 5%, so that the approximation of constant a, b and c should be valid.

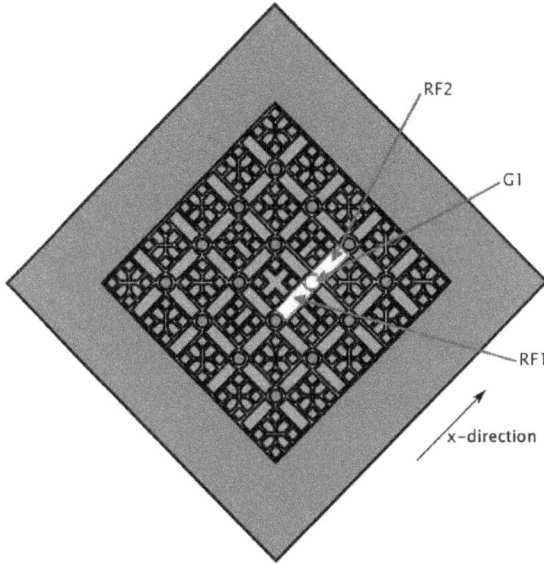

Figure 6.6: Folsom trap RF displacement simulation setup. The electrodes RF1, RF2, and G1 are the relevant electrodes for computing the dipole and quadrupole coefficients around the trapping site electrode G1. The residual RF field from the rest of the RF electrodes (colored red), must also be taken into account.

6.2 RF Resonators

In order to create the relatively high voltage RF signal, which is of order 100 volts, necessary to drive ion trap electrodes, a resonant circuit is typically employed. Since most RF sources are $50\,\Omega$, so that standard coaxial cables can be used, without a resonator, the power to drive an electrode would be $P = V^2/R$ or several 100 W. Such an RF source would be technically difficult. Many ion traps employ a helical resonator [208], which can achieve a very high voltage gain (typically $\sim$50) and a loaded Q of several hundred. However they are typically $\sim$10 cm in diameter and $\sim$20 cm long at typical ion trap drive frequencies. Since they must be placed as close to the trap as possible, typically on the vacuum feed-through it would be space prohibitive to use more than just a few helical resonators. For the addressable 2D arrays several RF sources were required, so a smaller resonator was designed and tested based upon toroidal inductors.

In the following subsections, first a simple model of an LC resonator is presented (6.2.1). Then the results of testing are real resonator based upon toroidal inductors is given (6.2.2). A technical complication is described, which arises when multiple resonators drive electrodes, which have some capacitive coupling to each other, are driven separately (6.2.3). And finally, a solution to this problem is given (6.2.4).

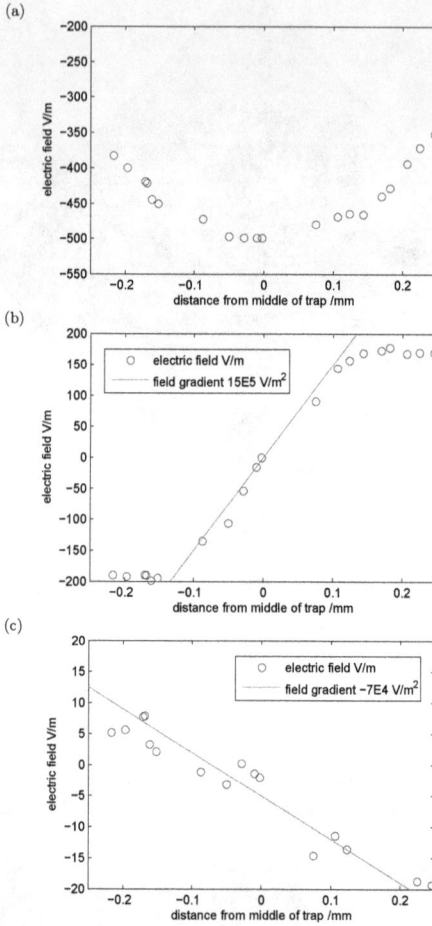

Figure 6.7: The electric fields as a function of the x position around the center of a trapping site for the various applied voltage configurations. A procedure analogous to that given in section 3.5.2 is used, so that the simulations are run as follows: For the dipole-like electrode configuration figure (a), corresponding to coefficient a, the two near RF electrodes have a voltage of $+1\,\text{V}$ and $-1\,\text{V}$ while the rest of the array is at ground. For the quadrupole-like configuration figure (b), corresponding to coefficient b, the two nearby electrodes have a voltage of $+1\,\text{V}$, while the nearby ground electrode and ground plane are set at $0\,\text{V}$. For the residual array RF figure (c), term c, all the RF electrodes, except for those neighboring the trapping site are set to $+1\,\text{V}$, while all other electrodes, including the neighboring RF electrodes, are set at $-1/2\,\text{V}$. In order to compute the quadrupole coefficients, a line has been fitted to the simulation data, so that the gradient of the electric field can be seen. The extra field has the effect of producing a quadrupole like field which is only about 5% compared to the adjustable RF electrode's quadrupole term. For these simulations, a ground plane sits $7\,\text{mm}$ above the surface of the trap.

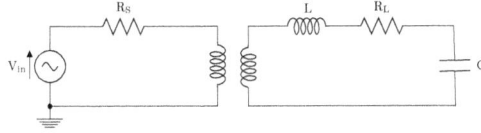

Figure 6.8: Schematic representation of a tank resonator, which uses a perfect transformer to match the resonator's resistance R_L on resonance to that of the source impedance R_S of the voltage source V_{in}.

6.2.1 LC Resonator Model

A common way of viewing the operation of these resonators is as an impedance matching network which matches the impedance of the RF source to the impedance of the electrode [209]. Since the electrode and associated wiring is essentially a capacitive load, the RF power necessary to drive the electrodes is only a function of how good of a resonator can be built (a better resonator requires less power). This allows for less RF power to be used driving the trap, while it also acts as a band pass filter, which can reduce voltage noise on the electrodes.

A simple model of a resonator, with a perfect matching network is given in figure 6.8. The signal source with a source impedance of R_S is perfectly matched at the resonant frequency to a tank resonator composed of an inductor L, some resistance R_L, and a capacitor C. On resonance, the impedances of the inductor and capacitor cancel each other. All the resonator losses are lumped into the resistor R_L. At steady state, the power supplied from the signal source $P_S = V_{in}^2/R_S$ is lost in the resonator's resistor R_L. The power lost in the resistor R_L is $I^2 R_L$, where I is the RMS current flowing through the resistor. The RMS current I_{RMS} is equal to $V_{out}\Omega_{rf}C$, where V_{out} is the RMS voltage on the capacitor. Since the resonator is perfectly matched to the source, one can equate the power in and the power lost to arrive at an expression for the voltage gain G of the resonator

$$G = \sqrt{\frac{Q}{\Omega_{rf}CR_S}}. \tag{6.5}$$

From the above equation, one can then see that if is desirable to minimize the RF source's power, but the trap drive frequency and voltage are specified, then it is desirable to maximize the quality factor Q, minimize the capacitance C and if it is possible, to minimize the source impedance R_S.

6.2.2 LC Resonator Results

Shown in figure 6.9 is a tank resonator, which uses two capacitors to match the source to the resonator. An example of this type of tank resonator was built and installed on an existing ion trap experiment and $^{40}Ca^+$ ions were loaded. The trap was a linear quadrupole ion trap (QIT) with planar trap electrodes and an ion-electrode spacing of 475 µm. The drive frequency was 10.5 MHz. A 5 W amplifier[5] was used to drive the resonator. The elements for the circuit were

---

[5]Minicircuits ZHL-5W-1

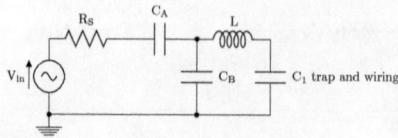

Figure 6.9: Schematic representation of a tank resonator, which uses an impedance matching network of two capacitors C_A and C_B, to match the resonator's impedance to that of the source impedance R_S of the voltage source V_{in}.

a 4.7 µH inductor (API Delevan 4470-09F) with a measured unloaded Q=84 at 10.5 MHz, and RF capacitors, C_A=220 pF, and C_B=820 pF. The capacitors were chosen to match the source impedance to the load impedance, so as to maximize the voltage gain. The measured capacitance of the trap, associated wiring and vacuum feed-through was 47 pF. A capacitive divider was used at the output of the resonator to measure the voltage of the resonator. The resonator gave a voltage gain of 22.5 and a measured, loaded Q of 51 (resonant frequency divided by FWHM of bandwidth). This circuit could output 1000 V peak to peak continuously.

Though this resonator loaded ions in a linear QIT, an improvement to the matching network of C_A and C_B would be to provide a DC path to either ground or a DC voltage source. This could be done in various ways, for example by adding an RF choke (a large inductor) at the junction of C_A and C_B to a DC source. Alternatively, one could eliminate C_A and use an inductor to make the match instead [210, 211].

6.2.3 Capacitive Coupling between Two Resonators

The simple, impedance-matched tank circuit described above works if only one radio frequency drive is required. If it is necessary to adjust one electrode to a different RF voltage amplitude than that of the others, then it might seem that a scheme involving two resonators and two variable sinusoidal sources could be employed. However, a problem surfaces because of the coupling between the two resonators.

Figure 6.10 shows two resonators, like the resonator shown in figure 6.9, each driving their own capacitive load C_1 and C_2. These two resonators are weakly coupled because they drive electrodes which have a capacitive coupling, $C_{coupling}$, at the trap. When one electrode voltage is adjusted, it will affect the other resonator and both the phase and amplitude of the RF driving voltages will be affected. For example, if both resonators have the same output voltage phase and amplitude at node points 1 and 2 (see figure 6.10) then the coupling capacitor $C_{coupling}$ is essentially invisible to the system. However, if the second resonator is set so that node point 2 is at ground, then the capacitance that the first resonator needs to drive will increase by $C_{coupling}$. While this is typically very low ($\sim$0.1 pF) it is enough to significantly affect the phase and amplitude of the output of the resonators. To first order, the change in resonant frequency $(\delta\Omega_{rf}/\Omega_{rf}) \propto \delta C/2$, where δC is the fractional change in the capacitance of the resonator. Since the resonators might have a loaded Q of $\sim$50 and a load of no more than a few tens of pFs, a change of 0.1 pF would result in a phase shift of about 20°. The ion would not be stable with such a large mismatch of phase between the voltages on the RF electrodes.

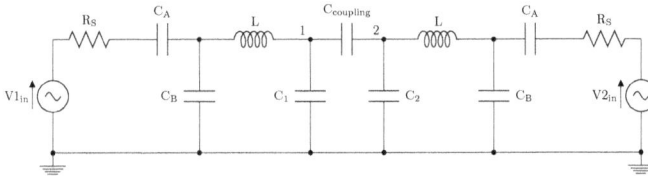

Figure 6.10: Schematic representation of two coupled tank resonators. The capacitor between the two trap electrodes $C_{coupling}$ is the capacitive coupling caused by the physical proximity of the electrodes on the ion trap. The voltage (amplitude and phase) at node 1 becomes dependent not just on the sinusoidal source $V1_{in}$, but also the voltage at node 2, leading to instabilities in the drive.

There are multiple ways to avoid this problem. One could do away with the resonators, but the required RF power ($\sim$200 W) to drive the ion trap becomes very expensive and technically difficult. The trap could also be operated with a low-Q resonator [193], but instabilities in the phase would likely still cause high micromotion and the RF power required ($\sim$10 W) makes scaling the system difficult. A third way, used here, is to actively lock the resonator.

6.2.4 Phase-Locked Resonator

A varactor diode as a variable capacitor can be used to compensate for the variable capacitance of the trap electrodes, which is caused by their capacitive coupling to other electrodes. Figure 6.11 shows a representation of a circuit to do this. The output of the resonator is measured with a capacitive divider (C_2 and C_3, typically 1 pF and 100 pF respectively) and the phase of this resonator is computed with a mixer (Analog Devices AD8302).

A resonator has a phase shift of 90° between the input and output when it is on resonance. As the input frequency is varied the phase will change with a linear slope around the resonance frequency. A mixer can be used to compare the phase of the output of the resonator, with the phase of the input of the resonator. The output of the mixer is then used as an error signal, which is sent to the varactor diode, so as to lock the resonator to a fixed phase. Care must be taken to protect the varactor diode from the high voltage output of the resonator. This can be done by using another capacitive divider (with C_V typically $\sim$2 pF) to lower the RF voltage the varactor diode sees. The output of the mixer must be be low-pass filtered, biased, and given the correct gain to properly drive the varactor diode. Care must also be taken to keep the varactor diode reverse biased (not shown in figure).

Initial tests of this phase-locked resonator allow it to compensate for $\sim$2 pF of capacitive coupling at 9.7 MHz with less than 1 degree of phase change utilizing a resonator with the same characteristics as given above in section 6.2.2. Without the phase lock, the resonator had a phase shift of $\sim$30 degrees. A full electronic schematic of the circuit used for operating Folsom is given in section A.

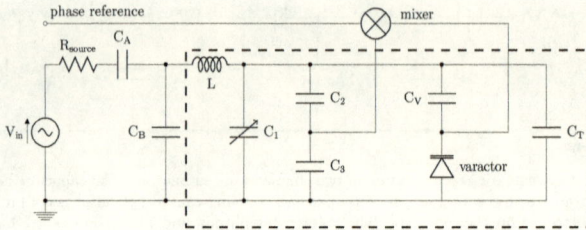

Figure 6.11: A schematic of the phase-locked RF resonator. A tank resonator is formed by the high Q inductor L and the capacitors to the right of it. The mixer locks the output phase of the resonator via the varactor diode.

6.3 Folsom Electrode Setup

To trap ions in Folsom, it needs to be connected to a set of RF and DC voltage sources. Figure 6.12 shows a false color image of the individually adjustable electrodes in the center of the array and their identifying number. The circular trapping site electrodes 1, 5, 13, and 20 are connected to a DC voltage source and the remaining numbered electrodes are connected to RF sources. Some of the electrodes have the same number, which indicates that they are internally connected to each other. Not shown are the outer trapping sites and the transparent conductive plane 7 mm above the trap, which are connected to the local ground.

Initial testing of the phase-locked resonators revealed that if the capacitive load of the tank resonator was varied by a few pF, it was possible to monitor the phase of the output of the resonator, and correct for this by using a bank of varactor diodes to compensate for the change in the load. Setting up several of these resonators for ion trapping was done as follows. Each resonator drove either a single, or group of, adjustable RF electrodes. Table 6.1 shows how each individually adjustable electrode, or set of electrodes, was connected to Folsom. The first column shows the trap electrode or group. For the RF electrodes, the capacitive load, voltage step-up and electrode numbers are shown in the remaining columns. For the circular DC trapping site electrodes, the electrode number is given in the final column.

Figure 6.13 shows the setup for each of the five RF sources, which were connected to the electrodes in Folsom. Phase-locked function generators were used to supply both a reference clock, along with the trap drive signal. The trap drive signal was amplified[6] and then passed through a directional coupler. Pickups in both the forward and reverse direction then went to an oscilloscope to measure both the forward and reverse power. A bias-T allows a DC voltage to be placed on top of the RF drive. The RF drive then went to a phase-locked resonator, as described in section 6.2.4. The phase-locked resonator used the reference clock to stabilize the phase of the RF drive.

Setting up multiple RF drives was done by starting at a configuration where all RF electrodes had the same voltage and phase, so that directional couplers could be used to tune the

---

[6]Minicircuits ZHL-2W

tank resonators and minimize any reflected RF power. However the situation is more complicated when the RF voltage on adjacent electrodes is not the same. Some amount of RF power is coupled capacitively from one adjacent RF electrode to another and then is transmitted up the line back through the resonator to the RF source, making it hard to use directional couplers to check whether the resonator is well matched. For this reason, an oscilloscope with capacitive pickup probes was also used to monitor the output phase and amplitude of each RF drive. RF drives with a large capacitive load of approximately 100 pF did not need the phase lock activated for the experiments. But lightly loaded (30 pF) RF drives required the phase-lock to keep the phase-walk to about 1° or less.

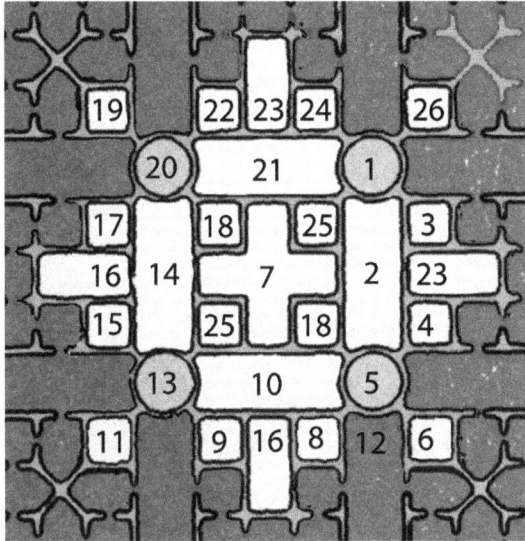

Figure 6.12: Numbering for Folsom's segmented electrodes is shown above. Some of the electrodes were wired together at the trap, so as to conserve the number of external connections necessary. The internally connected electrodes are shown here with the same number.

trap electrode or group	load / pF	voltage step-up	list of electrodes
adj 1	30	20	10
adj 2	30	20	14
main N	150	14	2,3,4,6,7,8,9
main E	150	14	11, 12, 14, 15, 16, 17, 18, 19
main W	180	13	21, 22, 23, 24, 25, 26
site A	n.a.	n.a.	13
site B	n.a.	n.a.	5
site C	n.a.	n.a.	1
site D	n.a.	n.a.	20

Table 6.1: Details of RF resonator setup and DC electrode drive for Folsom. The channel of the resonator, the load on the resonator and the voltage step-up are given. For the trapping sites A-D, the electrode number is given.

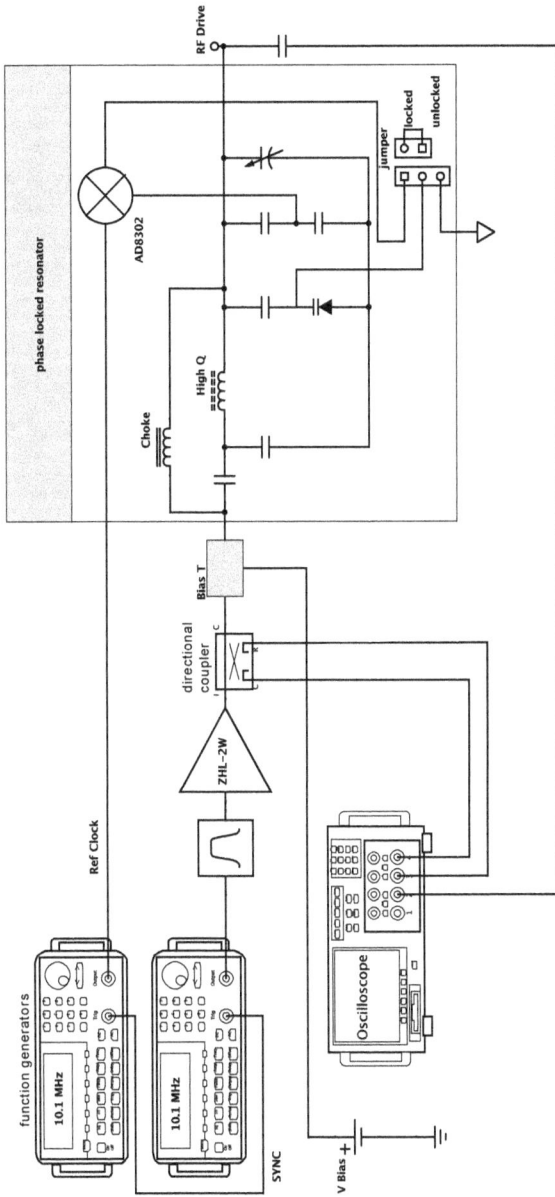

Figure 6.13: Setup for each of the five RF sources, which were connected to the electrodes in Folsom. Phase-locked function generators were used to supply both a reference clock, along with the trap drive signal. The trap drive signal was amplified and then passed through a directional coupler. Pickups in both the forward and reverse direction then went to an oscilloscope to measure both the forward and reverse power. A bias-T allows a DC voltage to be placed on top of the RF drive. The RF drive then went to a phase-locked resonator, as described in section 6.2.4. The phase-locked resonator used the reference clock to stabilize the phase of the RF drive.

6.4 Folsom Results

This section describes the main results of trapping and controlling ions in the ion trap "Folsom". First the results of trapping single ions in an outer site of the array are given (6.4.1). Then the results of trapping in a tunable region of the array are given (6.4.2). Next, the results of using adjustable RF to displace the ions in the tunable regime are shown (6.4.3). The method of micromotion minimization and frequency tuning using an adjustable RF electrode is then described (6.4.4). Next, the measurements of the trap frequency vs. the applied RF power are given (6.4.5). Finally the results of heating rate measurements via a recooling method are described (6.4.6).

6.4.1 Initial Trapping

After fabricating the ion trap (see sec. 6.1.2), the trap was assembled into an ultra-high-vacuum (UHV) chamber (see sec. 4.1). A calcium oven was used with photoionization lasers to create $^{40}Ca^+$ ions at the center of one of the outer trapping sites in the 2D array. Shown in figure 6.14 is a camera photo of a cloud of laser-cooled $^{40}Ca^+$ ions above one of the outer trapping sites in Folsom. The figure shows the electrode structure underneath the ion-cloud. Single ions were trapped in one of the outer trapping sites in the array using just a single RF trap drive. A secular frequency of 680 kHz was measured with a trap drive of 10.7 MHz at a voltage amplitude of 74 V_{rms}.

In order to measure the trap frequency, the following method was used. A sinusoidal voltage is applied to a nearby electrode in the array. And the frequency of the applied signal, with a minimal voltage that would heat the ions resonantly out of the trap, is then the trap frequency. Because the cooling laser beams were directed parallel to the surface of the trap, it was not possible to efficiently cool the motion of the ion normal to the trap's surface. A possible method to cool in this direction is described below in section 6.5.4. The measured secular frequency is that of the direction normal to the trap surface. For spherical ion traps this is normally called the axial or z-direction. The lifetime of a single ion with the cooling lasers turned on was approximately 10 minutes, allowing measurements to be made with the photoionization lasers turned off. However, the calcium oven, since it had a warm-up time of several minutes could not be turned off. The time required to load an ion depended on the laser parameters, and would range from a few minutes to just a few seconds.

A heating rate was approximated by an *ion-loss* method. The cooling lasers were turned off for increasing lengths of time. The time at which the ion-loss rate was approximately 50% was ∼ 5 seconds. Simulations of the ion trap give a trapping depth of ∼ 0.1 eV at the above operating parameters. Assuming a thermal distribution of the ion's motion, and calculating the expected loss-rate for the trap's simulated depth, a heating rate of ∼ 200 K/s was inferred. This also assumes that the ion starts from a temperature much less than the final temperature. It may be that the ion is quite hot in the direction normal to the surface of the trap (because it was not laser-cooled in this direction), and so this estimate can be considered an upper limit to the heating rate.

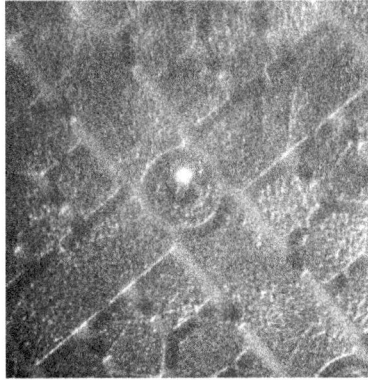

Figure 6.14: A cloud of $^{40}Ca^+$ ions loaded in Folsom over one of the outer trapping sites. The electrode structure (at the focus plane) can be seen below the ions.

6.4.2   Trapping in a Tunable-RF Region of Folsom

A novel RF drive, capable of driving adjacent RF electrodes with the same phase but variable amplitude was constructed (see sec. 6.2.4 for details). During the construction of this RF drive, the ability to load ions in the outer trapping site deteriorated and after several months it was not possible to load. After disassembling the experiment and rebuilding the trap with a new version of Folsom, it was possible to load ions again and perform measurements. A notable change to the experimental setup was that a transparent indium-tin-oxide (ITO) ground plane was installed roughly 7 mm above the surface of the trap array. Ions could be loaded $\sim 400\,\mu m$ above one of the inner 2×2 trapping sites, where the neighboring RF electrodes can be individually adjusted.

The trap had similar uncooled lifetimes as described above in section 6.4.1, but this time there were five high-voltage RF sources running in parallel to drive the different adjustable RF electrodes. Initial setup of the RF drive was done so as to place the same RF voltage amplitude and phase on all of the RF electrodes, mimicking a situation where the RF electrodes were simply connected to one another. The RF drive voltage was $\sim 75\,V_{rms}$ at 10.1 MHz. See section 6.2.4 for a description of the RF resonator and how it is used to generate the high voltage on the electrodes.

The voltage at the electrodes was then varied by changing the power applied to the RF resonator. All RF power data described in the following sections is in dB, relative to this initial situation where all electrodes have the same voltage. One important technical detail is that trapping ions now required that the circular trapping-site electrode had a negative voltage applied to it. At first this voltage was about -1 V, but over 1 year, the voltage required became more strongly negative; at times as much as approximately $-6\,V$. Figure 6.15 shows the progression of the required trapping voltage over time. In the outer trapping sites, the circular trapping electrode was simply wired to ground, and it was not possible to vary its voltage. If we had been able to apply a negative voltage to the outer trapping sites, it might have been possible to continue trapping. Various reasons for this required bias voltage are discussed below in section 6.5.1.

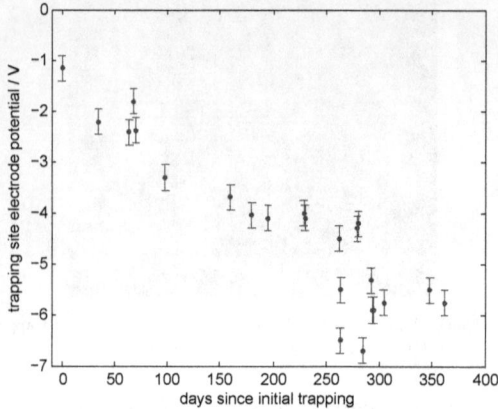

Figure 6.15: Above is the required voltage applied to the circular trapping-site electrode so that trapping was possible. Over a year of experimentation, the required voltage went from approximately -1 V to -6 V.

6.4.3   RF Displacement

One of the main features of using addressable RF electrodes, described in section 3.5.2, is that the ion(s) would be displaced by reducing the power to the individually adjustable RF electrode. The power to the electrode is proportional to the square of the RF voltage amplitude. As the electrode voltage is reduced, the electric field above the electrode falls, creating an asymmetry of the pseudopotential. The ion then moves towards the region of reduced electric field. In the limit of the variable RF electrode being at zero potential, the two nearest-neighbor traps essentially become a single, quasi-linear trap.

Figure 6.16a shows the electrode configuration for the RF displacement. To demonstrate displacement, the power applied to the y-adjustable RF electrode was varied. Figure 6.16b shows the ion's displacement as a function of the power applied to the adjustable electrode. The ion was displaced by 37 μm for a 5 dB reduction in the power. The expected RF displacement is also shown (solid line) from theory as per equation 6.4, with only the initial position as a fit to the data. RF displacement in the x direction was also performed with the addition of a second, adjustable, phase-locked RF drive applied to the x-adjustable electrode (see Fig. 6.16a). This allowed the ion to be displaced "around a corner" in the ion trap. The ion could be displaced approximately 40 μm in the either the x or y-direction, by reducing the applied power by $\sim$ 5 dB on the relevant electrode. This experiment could be repeated, however the lifetime of the ions was only a few seconds while performing RF displacement so that in order to do this experiment, ion loading was kept going. This meant that at times there was one, two or as many as four ions in the trap during RF displacement.

In general, if the adjustable electrode's phase is not exactly that of the other RF electrodes, excess micromotion at the trap drive frequency will occur [136, 167]. When the ions were displaced in the y-direction the phases of the five RF resonators were not locked. As the cooling laser was perpendicular to the direction of displacement, any micromotion in this direction did not interfere with laser cooling. Nonetheless, during displacement the cooled-ion lifetime was only a few seconds. This was possibly due to variations in the cooling power over the displacement range. For displacement in the x-direction a phase-locked resonator was used in order to minimize micromotion along the direction of the cooling beam.

6.4.4   Micromotion Minimization and Secular Frequency Measurement

In order to minimize the ion's micromotion and measure the trap's secular frequencies a method was used which recorded the time arrivals of the ion's fluorescence photons and compared these to the time varying trap drive signal. Fluorescence light from a single ion, at 397 nm, was measured by a photomultiplier tube (PMT) as described in the imaging setup (see sec. 4.4) and the resulting pulses were sent to a time-correlated single photon counter (TCSPC)[7]. The arrival times of the photons could then be analyzed in the frequency domain by taking a Fourier transform and inspecting the power spectral density. So that this could be done immediately while running the experiment, a computer program was written to interface with the TCSPC and the results were then analysed immediately and shown to the experiment operator. The DC voltages, bias voltages and RF electrode phases were then adjusted to minimize the micromotion. To measure the secular frequency, a nearby electrode was used to excite the ion's secular motion so that the Fourier analysis of the fluorescence photons could reveal the resulting secular motion peak [212]. Applying a weak voltage to a nearby electrode so as to excite the secular frequency is often called *tickling*.

In order to minimize the micromotion in both directions parallel to the surface of the trap, Doppler cooling lasers were set up in both directions. By varying the DC bias voltages on nearby electrodes, the ion could be placed closer to the RF null, reducing the micromotion peak. Micromotion minimization was done with application of one beam at a time as well as two beams at the same time. Because of the large number of experimental parameters, it was not possible to relate the size of the micromotion peak to a physical quantity. However, reduction of the micromotion peak allowed the micromotion to be minimized in a qualitative way.

Figure 6.17 shows a plot of the secular and micromotion peaks in the power spectral density of the time arrivals of the fluorescence photons. Most prominently in the figure, is the micromotion peak at the trap drive frequency of 10.1 MHz. The secular frequency peak can also be seen. For the conditions shown, the secular frequency ω_x was approximately $600 \times 2\pi$ kHz. Also visible is the micromotion sideband at approximately $10.1 - 0.6 = 9.5$ MHz. The appearance of the micromotion sideband is due to the excited secular motion of the ion mixing with the micromotion.

---

[7]Picoharp™ 300

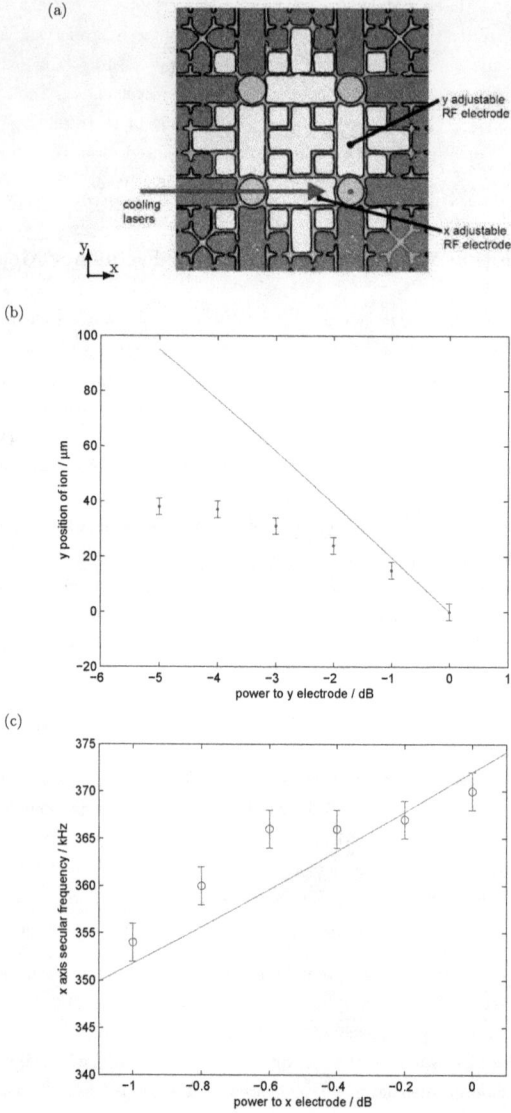

Figure 6.16: Results of adjusting the power to the adjustable electrodes. (a) the configuration of the laser-cooling and electrode adjustment. (b) the ion position versus the applied RF power. (c) the frequency of the ion trap versus the applied RF power. Also shown are the models of RF displacement and frequency tuning with a single parameter fit.

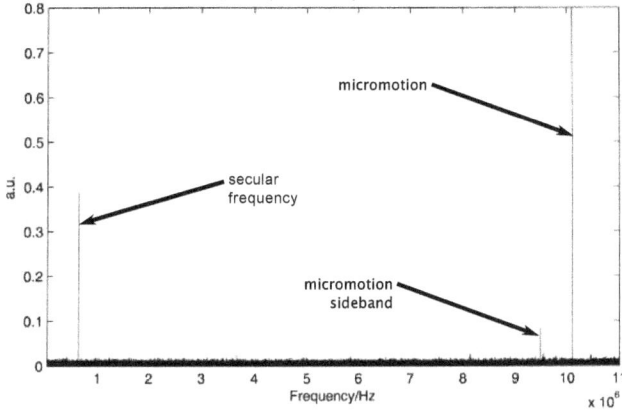

Figure 6.17: The secular and micromotion peaks in the power spectral density of the time arrivals of the fluorescence photons. Most prominently in the figure, is the micromotion peak at the trap drive frequency of 10.1 MHz. The secular frequency peak can also be seen. For the conditions shown, the secular frequency ω_x was approximately $600 \times 2\pi$ kHz. Also visible is the micromotion sideband at approximately $10.1 - 0.6 = 9.5$ MHz. The appearance of the micromotion sideband is due to the excited secular motion of the ion mixing with the micromotion

6.4.5 Secular-frequency Tuning

In addition to facilitating RF displacement, changing the power applied to the adjustable electrodes modifies the trap's secular frequency. Figure 6.16a shows the setup for adjusting and measuring the secular frequency via the second adjustable RF electrode and figure 6.16c shows the adjusted trap frequency (350 to 370 kHz) in the y direction as a function of the power applied to the second adjustable electrode. A model of frequency verses applied power is also shown (solid line) following equation 3.48, where the only free parameter is the initial unmodified trap frequency. The frequency was adjusted by about 15 kHz (4%) with a 1 dB change in the applied power. It should be noted that, in the absence of DC-voltages, the secular frequency of the radial direction is half that of the axial direction in point quadrupole ion traps with a cylindrically symmetric trapping potential [49]. The axial secular frequency in the absence of DC voltages had been measured to be approximately 700 kHz or about double the radial frequency. The difference can be accounted for by noting that a small negative DC voltage (-2V) was applied to the circular trapping site, which increased the radial frequencies at the expense of the axial frequency.

The tickling method by which the secular frequency was excited and thereby measured used a near-resonant Doppler-cooling beam. This necessitated well-compensated micromotion, so that the Doppler-cooling beam was not modulated too far into the higher order (especially blue) micromotion sidebands, which would reduce the effectiveness of the Doppler cooling beams. Consequently, tuning and measuring the trap frequency could only be done with the RF resonator phase-locked to a reference. In this way, excess micromotion due to a phase difference on the RF electrodes was minimized.

6.4.6 Heating rate

While the ion-loss method outlined above provided an upper-bound estimate on the ion heating rate, it was desirable to characterize the heating rate further. The ion-heating rate was too high to employ sideband resolved spectroscopy methods [73], so a fluorescence recooling method was used [134, 135]. By this method the motional state of the ion after a period without cooling is inferred by time-resolved measurement of the fluorescence dynamics as it is recooled. The general theory for measuring the heating rate with a recooling method is given in section 2.10.3.

The heating rate measurements were performed as follows. Ions were loaded by setting all bias and DC voltages to zero, except for the circular trapping-site electrode. Because of stray field drifts over time (see figure 6.15), it was necessary to apply approximately -5.3 V to the circular trapping-site electrode to trap ions. This large negative voltage compensated for the stray field at the trap center, but modified the axial and radial trap frequencies as follows. The measured radial trap frequency increased from about 300 kHz to 630 kHz. As the DC field increased the radial frequencies, it necessary reduced the axial frequency to below 100 kHz (see section 2.1.1). The exact value of the axial frequency varied from day to day due to the changing stray fields in the trap but stayed below 100 kHz.

Figure 6.18a shows the fluorescence rate during recooling of an ion after it was allowed to heat for 24 ms. The data is an average of 1000 experimental runs. The detuning, Δ, of the 397 nm cooling laser was roughly $15(5) \times 2\pi$ MHz, where the natural linewidth is $\Gamma = 21 \times 2\pi$ MHz and the saturation parameter from the measured laser intensity was $\sigma = 10 \pm 2$. Because the axial frequency is so low compared to the radial frequency, heating primarily occurred in this direction. By modelling the recooling process, the average temperature in the axial direction was estimated. The cooling lasers only had a small overlap ($\sim 6°$) with the axial motion of the ion, so the cooling was not very efficient in this direction. Similar measurements were carried out for different heating times, shown in Fig. 6.18b. Fitting these points with a line gives a heating rate of approximately 2.5 K/ms. The uncooled lifetime of ions in the trap had degraded over time so that, when these heating-rate measurements were performed, the uncooled ion lifetime was approximately 100 ms.

In implementing the recooling model [134, 135], the following details of the calculation are given. For the corresponding energy average appearing in the recooling model, the energies $E \in [E_D, E_{max}]$ are included. E_D is the Doppler limited minimal energy and E_{max} is a truncation value chosen to be high enough such that the omitted higher energy terms in the average do not alter the result of the fit routine. A value of $E_{max} \approx 20\overline{E}$ was sufficiently high, where E_{max} is adapted according to the size of the fit parameter $\overline{E}$.

The original recooling model, described in section 2.10.3, did not include any ion heating during recooling, and it was assumed that any heating during the recooling could be neglected. The rate of change of the normalized energy $d\epsilon/d\tau$ of the ion given by Wesenberg et al. [135] is

$$\frac{d\epsilon}{d\tau} = \frac{1}{2\sqrt{\epsilon r}}(\text{Re}(Z) + \delta \text{Im}(Z)), \qquad (6.6)$$

(a)

(b)

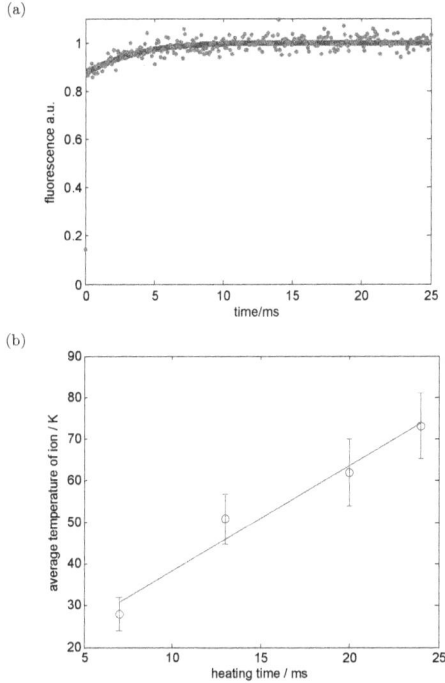

Figure 6.18: Heating-rate measurements using the recooling method. (a) fluorescence rate as a function of time during recooling (point data) and the recooling-model fit (solid line), after allowing the ion to heat for 24 ms. (b) the average energy of the ion, given by the recooling model, as a function of the heating time. A least-squares fit (dotted line) gives a heating rate of approximately 2.5 K/ms

where $Z = i/(\sqrt{1 - (\delta + 1)^2/4\epsilon r}$, τ is the normalized time, δ is the laser detuning, and r is the recoil parameter.

Our initial results indicated a heating rate of order $10^6 - 10^7$ phonon/s. Because of this very high heating rate, there was some concern that the recooling model might be invalid, since it did not include ion heating during the recooling process itself. By adding the normalized heating rate $d\epsilon_h/d\tau$ to the normalized energy trajectory (eq. 6.6), the model could be checked. Since the normalized heating rate $d\epsilon_h/d\tau$ is the value that the model is giving us, the model must be run iteratively so as to make sure that the model is consistent with itself. For calcium ions with a trapping frequency of roughly 1 MHz, a saturation parameter of about 1, detuning of 10 MHz, and a heating rate as high as 10^7 phonons/s, this caused the recooling fit to vary negligibly.

6.5   Folsom Discussion

Below we consider the results of trapping in Folsom and what this might mean for 2D arrays of ion traps. First, the negative voltage required at the trapping site is discussed (6.5.1). Then the RF displacement and whether the RF null is genuinely altered by the reduction of the applied RF power is considered (6.5.2). Next, the adjustment of the trap's secular frequency by modification of the applied RF power on the adjustable electrode is reviewed (6.5.3). Finally, the heating-rate measured in the array is discussed (6.5.4).

6.5.1   Trapping

The results above indicate that ions can be trapped in an array such as Folsom so long as it is possible to apply a static voltage to the circular trapping site electrode. Personal communication with other ion-trappers suggests that this negative voltage is often necessary for a planar electrode traps [213] and may be caused by contamination on the electrodes [214–218]. It will be important for this architecture to minimize any required DC voltage underneath the trapping site so that RF displacement is effective. Whatever caused the required negative trapping-site voltage, it became worse with time. It may be that the negative voltage was necessary because of calcium deposited from the oven over time, or perhaps from the copper of the electrodes diffusing through the gold plating over time or even growth of field emission tips. Future versions of 2D arrays will need to overcome this issue. One additional external connection that could be made, would be to the transparent ground plane above the trap. This would allow a vertical electric field to be applied across the whole array.

6.5.2   RF Displacement

The RF displacement described above, shows that by varying the power to the adjustable electrode, the location of the ion in the trap could be moved. The location of the pseudopotential was altered generally in accordance with the simulations in section 6.1.3. However, the distance that the ion traveled was about 40µm, whereas the expected distance was about 90µm from the simulations (see fig. 6.16b). There are at least two possibilities. One is that the stray field compensation did not compensate for the curvature of the stray field. Another is that as the RF power was lowered, the required stray field changed. i.e. That the stray field is a function of the applied RF power.

There could be concern that some other effect was moving the ions as the variable RF electrode's power was reduced. One possibility is that, as the pseudopotential became weaker, some static voltage then pulled the ion in the direction of the variable RF electrode. This is unlikely, since at the time we performed the RF displacement experiment described, it was required to apply approximately -2 V to the circular trapping-site electrode. All other static voltages were zero in the system, so it would be expected that the static electric field in the trap would be pulling the ion away from the variable RF electrode. Moreover, as the power was reduced below -6 dB, the fact that the ion reversed direction back towards the circular trapping-site electrode suggests strongly that there were no static electric fields pulling the ion towards the variable RF electrode. Our interpretation of this is as follows. As the RF power was reduced below -6 dB, the pseudopotential, parallel to the surface of the trap and in the direction of the variable electrode, became so weak that the negative static voltage attracted the ion back towards the circular trapping-site electrode. In other words, the spherical trap had become so weak parallel to the surface in the direction of the trap of the variable RF electrode, that it was essentially a planar-electrode linear quadrupole ion trap (QIT), where the circular trapping-site electrodes, on either side of the adjustable electrode, acted as as an end-cap electrode, albeit with a negative voltage on it.

6.5.3 Frequency Tuning

The measurement of the RF frequency tuning described above required the use of a phase-locked resonator so that the adjustment of the RF drive did not induce excess micromotion, which would reduce the efficiency of the laser cooling. The range over which the frequency was adjusted and measured was about 5% of the secular frequency, with a 3% change expected from equation 3.48. However, during the RF displacement experiments, the RF power was reduced by 68% (5 dB), so that the voltage was reduced by 43%. It would be expected that the frequency would have been reduced by about 80 kHz (see eq. 3.48) during the RF displacement. However, measuring this secular frequency change remains a technical challenge due to the varying Doppler cooling power, and varying stray field compensation requirements during shuttling.

6.5.4 Heating Rate

The heating rate measurement of order 10^7 phonons/s suggests an electric-field noise at the ion of roughly 10^{-7} V^2/m^{2}Hz. Such a high level of electric-field noise or ion heating is rare, but not unheard of [219]. Since the ion was only cooled along one axis, the ion was quite hot in the directions perpendicular to the cooling beam. However, because these directions are normal to the cooling beam, the thermal motion would have no Doppler shift in the direction of the cooling laser, and should not introduce any error into the heating rate estimate given above. Additionally, any micromotion of the ion, normal to the surface of the trap, should not cause any error in the above heating rate estimate. Such micromotion, perpendicular to the surface of the trap, is to be expected from the static voltage on the circular trapping-site electrode. Moreover, the order of magnitude of the heating-rate, due to the correlation with the ion-loss method, suggests that the heating rate really is this high. Clearly, this trap is very hot when compared to published rates of similar sized ion traps and it would never be used for quantum information experiments. This trap is made from a gold-plated-on-copper PCB. We believe this to be the first published heating rate of a trap made from such technology. While relatively quick and easy to fabricate, it may be that this technology produces a trap with far too high of a heating rate.

Compared to other ion traps with comparable ion-electrode distances, the heating rate in Folsom is 10-100 times higher [125]. But while the heating rate is very high, it does not appear to be caused by the operation of multiple RF electrodes, since the heating rate is also high with just one RF drive; i.e. there were similar uncooled ion lifetimes, of a few seconds, with just one RF drive and with multiple RF drives. It is possible that the high heating rate is caused by some surface mechanism. The surface is rough ($\sim$1 µm RMS) and the PCB's copper electrodes, had after some time diffused through the electroplated gold layer, leaving a gold surface with patches of copper. Small electron field emission currents ($\sim$ 1 µA) were measured, but operation of the trap drive at powers low enough to completely eliminate the emission was not possible. Applying a high voltage source to try and burn-out any field emission tips was also considered too risky. In the future, any 2D array of ion traps would have to be fabricated more carefully so as to reduce the heating rate.

The large heating rate in the direction normal to the trap surface, as well as the poor cooling ability in this direction, raises the question of how to cool ions in a 2D array in the direction normal to its surface. Lasers normal to the 2D array would generally require access holes in the array and tilting the motional axes of the ion would introduce an asymmetry to the whole array, which would add technical complexity to many other operations. Gorman et al. [220] have used a weak electric-field excitation to parametrically couple the normal and parallel motions of an ion in a surface electrode trap, so as to cool it to the ground state of motion and measure the heating in the direction normal to the array. This could also be used during normal Doppler cooling (or recooling) to increase the cooling efficacy in this direction.

Chapter 7

Microscopic Ion Trap Arrays

In this chapter some of the work describing 2D arrays of microscopic ion traps, where coherent interactions between the ions would theoretically be possible, is given. Two-dimensional arrays of trapped atomic ions made with planar electrodes require a design where islands of RF electrodes are separately driven from DC electrodes. This requires some sort of underlying 3D circuit structure so that the electrodes can be connected to voltage sources. The ion trap array Folsom used a printed circuit board and vias to allow the different electrodes to be connected. For the two designs below, vias were a critical part of their design.

The rest of this chapter is laid out as follows. First in section 7.1, the microscopic array "Ziegelstadl" is described, where a multi-layer lithographic method was used to build a 4×4 microtrap array on one side of a dielectric substrate. Then in section 7.2, preliminary steps are described which would allow through-substrate vias so as to build a microtrap array, which might be scalable to a thousand trapping sites.

7.1 Ion Trap Array "Ziegelstadl"

In order for ions in separate harmonic traps to coherently exchange quantum motion, it is desirable to get them within 10s of microns of each other [141, 142], while keeping the heating rate low. This requires miniaturizing a 2D array like Folsom, while using materials and techniques known to give low heating rates. This trap array is called "Ziegelstadl".

The rest of this section first describes the physical layout of Ziegelstadl and the results of building and testing this microscopic array (7.1.1). Then the RF resonators for this trap are described (7.1.2). Next, the details of the vacuum setup and a custom feedthrough for this trap are given (7.1.3). Finally, the delivery of the laser light to the array is described (7.1.4).

7.1.1 Ziegelstadl Design

In figure 7.1 is an image of a miniature version of a 2D array of ion traps. The distance between the traps is 100 μm to allow for loading and cooling of the whole array with a sheet of laser light. The electrical connections to the trap electrodes are made on the sides with gold bonds. Two layers of conductive traces with vias allow for the connections on the side to reach the inside of the array so as to drive the various electrodes. The adjustable RF electrodes between the trapping sites would allow for the ions to approach a distance of approximately 60 μm during addressing while maintaining a trapping frequency of at least 100 kHz. The height of the ion above the trap surface is roughly 50 μm. A trap drive frequency of 100 MHz with a 100 V_{rms} amplitude trap drive, would give a secular frequency of ~ 10 MHz with a trapping potential of ~ 0.1 eV.

A few different versions of Ziegelstadl were built[1] and tested. The details of the fabrication of Ziegelstadl are given in appendix F. The general structure of Ziegelstadl is as follows. A three layer structure is built onto one side of a wafer, which consists of a bottom metal layer, an insulating layer and a top trapping-electrode layer. The insulating layer is made from one micron of silica dielectric and should be capable of withstanding several hundred volts before dielectric breakdown [221]. The two metal layers are interconnected with vias, so that the segmented inner electrodes can be connected to an independent voltage source. The technical demands of building such an 2D array of microscopic traps are still being solved (see app. F) at the time of writing this thesis. However, electrical tests have been partially successful [222] and efforts are underway to install the array into a vacuum setup and trap ions.

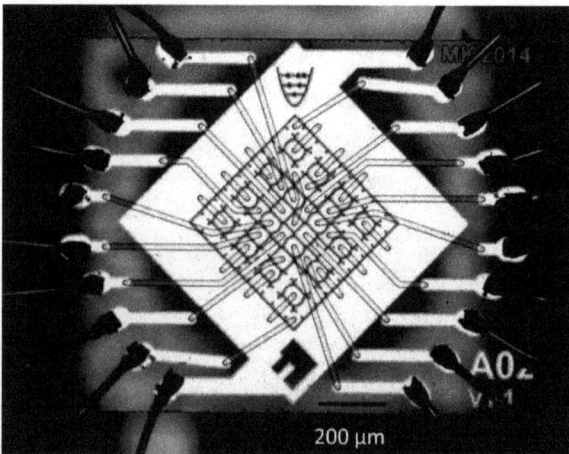

Figure 7.1: Above is a microscope image of a miniaturized 2D array of ion traps, called *Ziegelstadl* currently being tested. Here it is shown mounted and wire bonded to a filter board.

---

[1]fabrication primarily at Fachhochschule Vorarlberg by Prof. Johannes Edlinger's Microtechnology group

7.1.2 RF Resonators

RF resonators for Ziegelstadl were developed to run at frequencies of 60-100 MHz [222] so as to maintain the trapping parameter q_z to less than approximately $1/3$ (as it was in Folsom) while keeping the trap depth approximately constant. This was done so as to keep the micromotion no higher than already tested.

7.1.3 Vacuum Feedthrough and Chamber

In order to provide a low capacitance pathway so as to drive Ziegelstadl's RF electrodes with approximately 100 V and a frequency between 60-100 MHz, a custom fabricated vacuum feedthrough was developed [222]. The feedthrough is fabricated from a 3.1 mm thick printed circuit board made from Rogers™ 4350b laminate. A copper ring, which holds the circuit board and is sealed to it with a solder connection also acts as a Conflat™ style sealing ring for the UHV vacuum system. After bakeout at 130° C the pressure inside the chamber was measured to be of order 10^{-10} mbar. However, the permeable nature of the feedthrough material caused the pressure to slowly rise to roughly 10^{-9} mbar over several months. This could perhaps be alleviated by building the PCB feedthrough with more copper layers.

7.1.4 Laser Beam Setup

In order to illuminate the entire 2D array for laser cooling, without too much scatter from the surface of the trap, a sheet of laser light was required. Both the 397 and 866 nm laser optics needed to form a highly elliptical beam of light. Using cylindrical lenses, a sheet of laser light could be made 5µm thick and 200µm wide.

Photoionization was provided by a focussed 422 nm beam of laser light. However, an integrated LED light source could be used to provide 375-380 nm light since coherence is not required for the transition to the continuum [81].

7.2 Ion Trap Array "Matrix"

For 2D arrays of trapped ions with as many as 40×40 trapping sites, the basic architecture of Folsom or Ziegelstadl could be scaled-up; provided that the adjustable electrodes could somehow be connectorized. For Folsom, the electrodes were connected with through-substrate vias to traces on the back side of the printed circuit board. These traces were then connectorized with crimpable pins. For Ziegelstadl, a multi-layer micro-fabrication process created vias between the top layer electrodes and the bottom layer traces. These traces then fanned-out to pads, which could be connectorized with gold bonds. These gold bonds then connected to a filter board.

Unfortunately, the fabrication process used for Ziegelstadl will not allow for many more connections, as there is a limit to how many inner vias can be fanned-out on just a single lower layer. Consider that n is the number of sites of one side of the 2D array. The number of inner adjustable electrodes then scales with $n \times n$, whereas the length of the sides which allow for fanning-out of the traces scales only as n. What is needed is a three-dimensional connectorization process so that the inner adjustable electrodes can be connectorized. Modern micro-electronics with so-called flip-chip connects, use a three dimensional silicon fabrication process utilizing many connection layers. If 2D arrays of miniature ion traps are to be scaled-up, some form of 3D interconnection process will be needed.

7.2.1 Electroplating Gold Vias

One way would be to create through-chip vias, so that gold bonding on the reverse side could be used to to finalize the interconnects. Such a design, called *Matrix* is shown in figure 7.2. Figure 7.2a shows the front with its segmented adjustable electrode structure. The design of Matrix is a scaled-up version of Ziegelstadl's electrode structure. Figure 7.2b shows gold bonding pads on the rear of the 2D array. Through substrate vias connect the electrodes on the front to the pads on the back. In order to test the feasibility of fabricating through-chip vias with

(a)   (b)

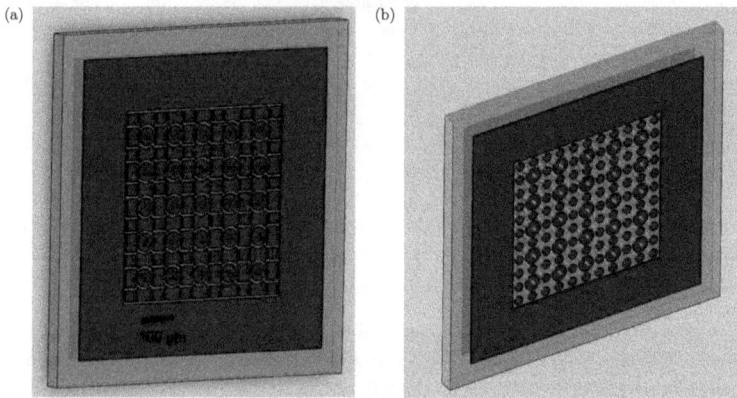

Figure 7.2: The 2D array design Matrix has a 5×5 array of addressable ion traps. (a) the view showing the segmented adjustable electrode and (b) the view showing the pads for gold bonding.

electroplating, a wafer was designed with 18×18 trap arrays. The method to metalize the vias was inspired from an expired patent [223]. A small run of 50 mm x 50 mm x 125 μm wafers were laser drilled with 20 μm holes for the vias[2]. One side of the substrate was then metalized with 50 nm of titanium and 100 nm of gold. One side of the substrate was then covered with gold leaf. After fixing the substrate to a fixture and gold bonding to the metalized substrate, the gold leaf was then sealed with photo-resist. A sulfite based gold-plating solution[3] was used to plate through the substrate. Results of the plating are shown in figure 7.3, where the laser drilled holes are shown in figure 7.3a and the results of the through plating are shown in figure 7.3b.

---

[2]laser drilling by 3D Micromac AG
[3]Transene TSG-250

The results indicated that after a planarization process, the electrodes could be patterned using standard photolithographic techniques.

Unfortunately, the yield of this process was quite low. Of the 324 trap arrays on the substrate, only about 3 arrays could be plated to look like that in figure 7.3b. The rest suffered from various defects. A large number of the trap arrays were filled with the photoresist because the gold leaf was porous. This could be mitigated by using a thicker electrode. If it were a gold electrode, this would be quite expensive, and if it were copper or silver, there is some concern that oxides might build up and it would be difficult to clean the surface prior to gold plating. Another difficulty is that the trap arrays on the edges of the substrate plated much faster than at the center of the substrate due to field line concentration between the cathode and anode in the plating solution. For these reasons, it appears that electro-plating would work, but a fair bit of work must be done to improve the yield before further processing steps could even be attempted.

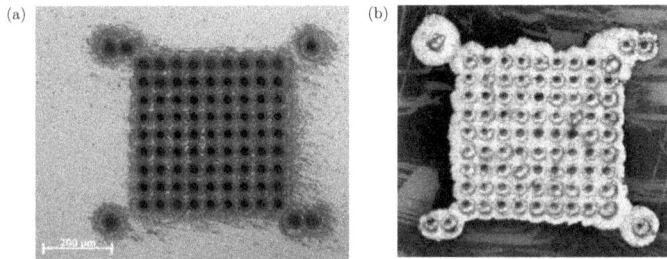

Figure 7.3: Steps showing Matrix fabrication. A laser drilled alumina substrate (a) is metalized with gold electro-plating so that vias form connections through the substrate (b).

7.2.2 Electroless Gold Vias

Because the electro-plating process produced such a low yield, another process to make gold vias in alumina substrates was tried using an electroless gold plating technique [224] utilizing a $Na_3Au(SO_3)_2$ solution[4]. This process involved four steps using chemical reactions. The first step uses silane to deposit a SiOH group on the alumina surface to activate it. The second deposits tin on the surface, a silver solution then precipitates to form a monolayer of silver over the tin, and finally the electro-less gold plating solution is used.

The original protocol was developed for producing gold nanotubes in anodic aluminum oxide (AAO) substrates. The technology of producing and metalizing such AAO substrates has since advanced. It is now possible to produce lithographically selective 3D nanotubes or vias in AAO layers [225, 226].

Initial results of electroless plating in nanometer sized holes with AAO substrates, as well as laser-drilled alumina substrates shows promise. The results of plating the laser-drilled substrates are shown in figure 7.4.

---

[4]available from Technic Inc., Cranston, RI: Oromerse SO Part B

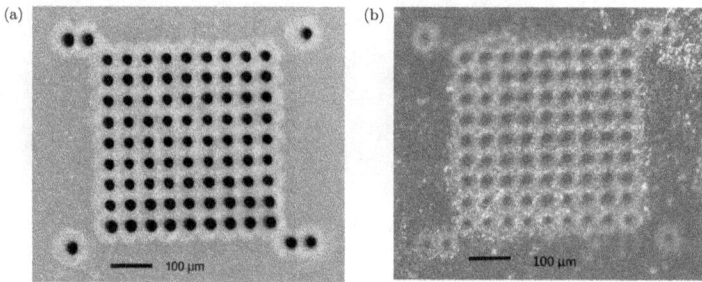

Figure 7.4: Steps showing Matrix fabrication, 2nd attempt. A laser drilled alumina substrate (a) before metalization (b) after gold metalization using an electro-less gold plating process.

Chapter 8

Outlook and Conclusions

In the preceding chapters, 2D arrays of ion traps have been simulated, designed and tested. The necessary theoretical background to understand their operations has been given. They appear to behave as predicted albeit with a few surprises. The segmentation of the RF electrodes, so that the time varying electrical potential on individual RF electrodes can be adjusted, has the desired effects of RF displacement and secular frequency tuning. This suggests that a 2D array of many ion traps could be successfully addressed electronically for large scale integration. The necessary electronics, while more complicated than the usual helical resonators of ion traps, work well and the electronics to lock the phase of the adjustable RF work properly.

That said, there remain several areas of concern as to whether this technology will perform well enough so as to allow for practical quantum computing. The motional heating rate of Folsom was so high that not only no coherent manipulation of the motional state is imaginable, but coherent manipulation of the electronic states seems to be out of reach. The heating rate is so high that the ion leaves the Lamb-Dicke regime in such a short time as to make sideband cooling impossible. The lifetime of the ion, even with the cooling laser on is only ten minutes and so many experiments are still out of reach. Many of these problems are likely material related. In order to test whether adjustable RF could be used with 2D arrays of ion traps, the short-cut of using a printed circuit board trap was used. In order to create the necessary geometry with materials known to have low heating rates, it will take a technology effort which involves the production of micro-vias. Some exploratory tests have been made to see if such geometry could be made in-house. It appears the answer is yes, but that it will take some time to get the process to work.

The voltage drift necessary to load the trap suggests strongly that the potential above the surface of the electrodes is being modified by the operation of the trap. Some process to reset the vacuum potentials above the trap surface should be integrated into the experiment. Other researchers have had success using ion-beams or strong lasers to clean the top layers of the trap [227–229]. These processes seem to eliminate both the static potentials above the surface and reduce the heating rate.

8.1 Scalability of 2D Arrays

The architecture proposed here requires of order $2N$ adjustable RF electrodes for N trapped ions, each of which could store a physical qubit. Because of the 2D array structure, the number of connections required scales linearly with N, but the width on a 2D circuit board to route the signals scales with $\sqrt{2N}$, so that the required number of connections will overwhelm the space on a 2D circuit quickly. This need not be the limit. If the power electronics necessary to drive the adjustable electrode is built into the 2D array, by integrating semiconductors into the micro-fabricated arrays, then an addressed digital communications bus could be used to command the adjustment of the RF amplitude. This suggests that the architecture could be made scalable without limit as far as the electrodes and electronics are concerned.

8.1.1 Optical Access and Imaging

If a global beam is used to drive transitions, while electronics are used for the addressing, the beam must be able to reach every ion. If the height from the surface of the array is fixed, while the array length is made longer, the objective used to bring the light to the ions will need to be brought closer so that the whole array remains within the Raleigh length of the beam. For an array trapping site spacing of approximately 100µm, an array of 100 by 100 ions (1 cm by 1 cm) should be possible. The characterization of each trapping site, must also be made scalable so that micromotion, the secular frequency and the splitting due to nearest neighbor coupling could all be characterized. For the characterization of a single trapping site, a PMT with a photon counter was used with a manually adjustable aperture. If instead of a manually adjustable spatial filter, an array of mirrors, such as those found on commercial digital projectors, were used, then each trapping site could be optically addressed with respect to the image the PMT sees. In this way a single PMT could be used dynamically to characterize each trapping site in the array.

8.1.2 Bootstrapping a 2D Array

In order to develop methods which allow a large number of trapping sites to be characterized, so that they can all be used as locations of high fidelity physical qubits, a number of methods should be developed to increase the efficiency of this characterization and bootstrapping. For instance, it should be possible to measure the secular frequency, the micromotion, and the heating rate by performing a g_2 measurement on the fluorescence light coming from the ions, if the right conditions are met. Some possibilities are that one could build and shake an optical lattice, tickle the ions electronically (as done in section 6.4.4), or increase the secular frequency so that it is as high as the Doppler cooling linewidth (making sidebands visible). A two-dimensional array of mirrors, such as is commonly found in digital projectors, could allow addressing of the image of the 2D array of ions to a single PMT. This would allow one to bootstrap the operation of a large array, so that its performance could be increased to the point where sideband resolved transitions on the qubit transition could be observed and further improved.

8.1.3 Adjustable RF

One aspect of the demonstrated method for scaling up 2D arrays of point ion traps has been the use of phase-locked adjustable RF voltages to displace ions in neighboring traps toward each other to increase their coupling addressably. As the array is miniaturized, the required RF voltages and hence the trap depth can remain the same if the trap drive frequency is increased proportionally. However, controlling the phase of a high voltage source as the frequency is increased could be technically challenging. The method used for the trap array Folsom used a matched phase-locked tank resonator (see sec. 6.2). However there are other possibilities. One is use a balancing set of electrodes to compensate for any change in the load of the resonator [162]. Another possibility is to use an RF current source to drive the LC resonator near resonance [230]. This would then avoid having to directly stabilize the phase of the RF drive. A third possibility is to drive the electrodes directly. Technically, this could be made considerably easier to implement, if a digital drive were to be used to drive the RF electrodes [231–233].

8.1.4 Trap Optimization and Simulations

The simulations in this thesis were primarily performed using the finite element analysis electrostatic software COMSOL™. Such an approach allows for an arbitrary 3D geometry to be simulated. However, for large 2D arrays, such an approach requires an amount of memory and processing time which scales roughly as the cube of the size of the problem. For the Ziegelstadl simulations, about 32 GB (gigabytes) of RAM was necessary, along with more than ten minutes of simulation time on a state-of-the-art personal computer. Furthermore, such an approach solves for the electrostatic potential everywhere in space. The resulting electric field has sharp changes in its gradient, due to the granularity of the meshing. Using a Biot-Savard type law for electrostatics [234], surface traps made from an infinite plane with gapless electrodes can be simulated much more efficiently [235]. One can also include the effects of a ground plane above the surface [236]. Using such an approach, the geometry of the addressable arrays proposed here could be optimized such as has been done for static arrays of point ion traps [163].

8.1.5 Applications to Quantum Simulations

Trapped atomic ions in 2D arrays have dipole-dipole coupling. This could provide a powerful tool for building quantum simulators which would allow for a better understanding of other physical systems. The coupling of nearby ions in separate traps has similar dynamics to Förster coupling [237] or even the coherent energy transfer seen in photosynthesis [238].

8.2 Low Heating Rates

Heating of ions in ion traps is a correlated source of motional decoherence. As such, it is necessary to minimize this as much as possible. The use of cryogenic ion traps is one way of minimizing this sort of heating. Cryogenic ion traps have been able to show the lowest measured heating rate in planar surface electrode ion traps made from intrinsic silicon [170, 239]. An advantage of using silicon to fabricate ion traps is that the industrial techniques to fabricate exact microscopic structures are well established and allow for complex designs to be implemented. Other techniques have also been shown to significantly reduce the heating rate, such as cleaning the trap with an argon-ion beam [228, 240]. The technology for developing microscopic arrays of ion traps with a low heating rate seems to be well at hand.

8.3   Outlook for 2D Arrays

The low heating rates of ions in microscopic arrays of RF addressable cryogenic ion traps, made with silicon fabrication methods and microvia technology would allow large scale 2D arrays to be brought to a practical computational level. These improvements would involve serious research and development efforts, but the payoff would be huge and could enable the application of already existing quantum algorithms to solve otherwise intractable problems in finance, cryptography, and natural simulations.

Appendix A

RF Resonators

A.1   RF Resonator for Voltage Magnification

In order to drive capacitive (or inductive) loads with radio frequency at voltages (or currents) that would require too much power if terminated with $50\,\Omega$, it is common to use a tank resonator to reduce the amount of power required. Such a resonators have been used to drive EOMs in the lab 4.3, RF magnetics for driving hyperfine level transitions [241], or the ion trap's RF electrodes 6.2.

Figure A.1 shows the layout for such a resonator. The RF source plugs into connector P1. Since most commercially available RF sources are $50\,\Omega$, due to the impedance of commercially available coaxial cables, an impedance match is required. Capacitors C1* and C2*, match the impedance of the RF source to the tank resonator's impedance. For the case of driving a capacitive load such as trap electrodes or an EOM, the inductor LQ can be built by winding thick wire onto a low loss magnetic core[1], or an inert toroid such as air or PTFE.

A.2   Phase-Locked Resonator

The phase-locked resonator is a tank resonator, but with the ability to lock the phase of the output of the resonator with feedback loop techniques. Figure A.2 shows the schematic of the circuit of a single channel of the phase-lock resonator. Varactor diodes are used to change the capacitive load of the tank resonator. The circuit has over 100 discrete components. The output of the mixer (U1) is used to control the bias on the varactor diodes (VD*A, VD*B). The op-amps (U3 and U6) allow the potentiometers to set the control loop gains (R1, R11), as well as the input offset (R9) and the output offset (R8) of the feedback. Additionally the circuit uses mirrored varactors to minimize phase distortion of the output.

---

[1] Micrometals, Inc. or Fair-Rite Products Corp.

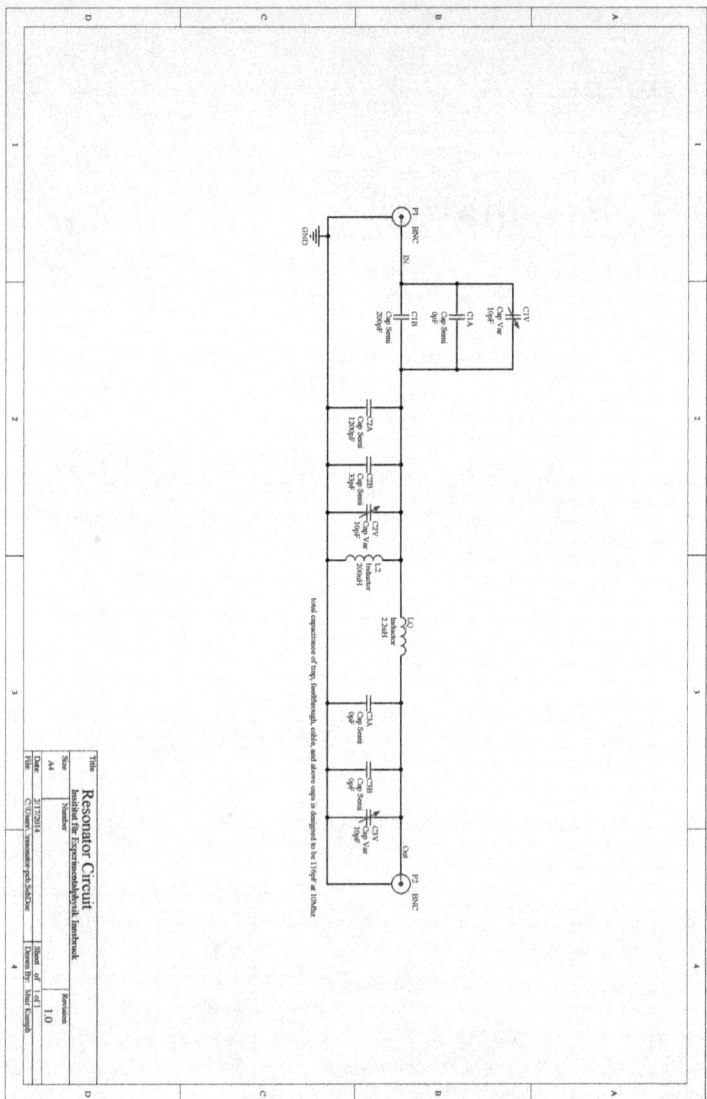

Figure A.1: RF resonator for voltage amplification. To reduce the power needed to drive capacitive loads, a tank resonator allows for the RF power to oscillate between the capacitive load C3* and the inductor LQ.

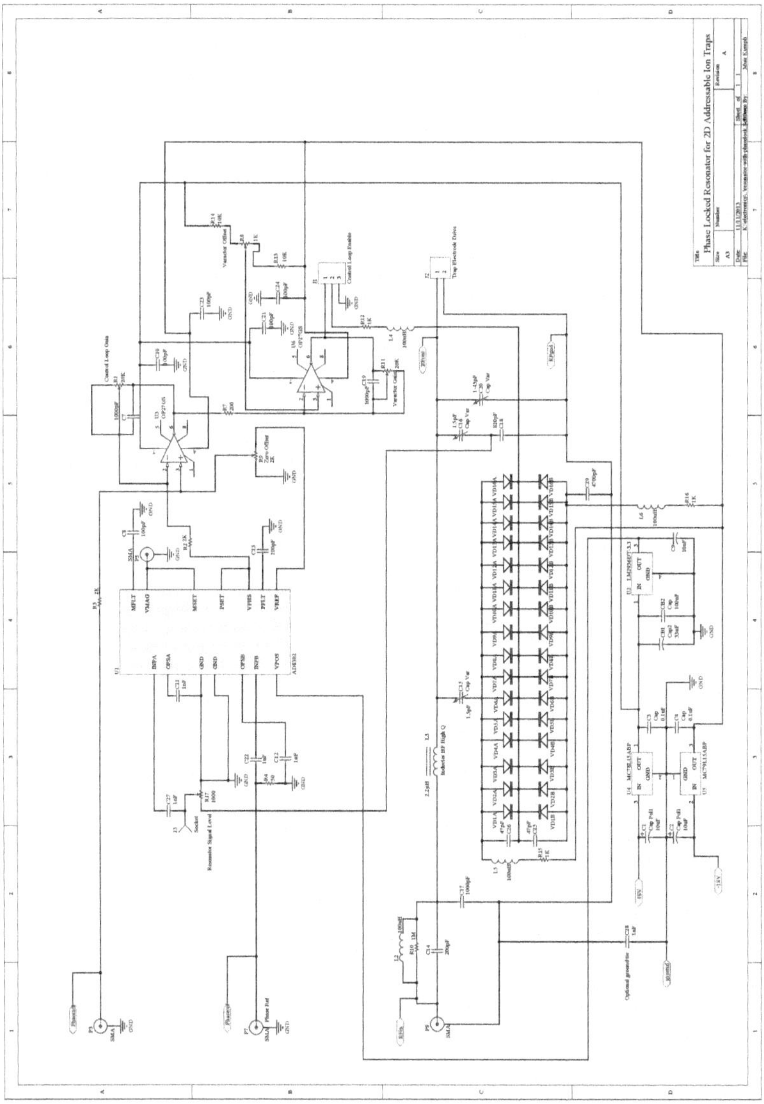

Figure A.2: RF phased lock resonator. The circuit controls the phase of the output of the resonator so that the phase is locked to a reference.

Appendix B

Control Electronics

A number of custom pieces of control electronics were developed in the course of setting up the laboratory for quantum optics experiments with trapped calcium ions. The schematics follow, so that a complete documentation is made of the experiment.

First shown in figure B.1 is a low resistance solid-state digital switch. It is used for driving moderately high current loads up to 7 A. This switch is used on the experiment to turn the calcium oven on. It is also used to control the solenoid based optical shutters, which allow for the zero'th order mode of the acoustic-optical-modulator (AOM) for the 397 nm laser beam to reach the ions for far detuned cooling. It is also used to control the shutter, which can block the photoionization lasers (375 and 422 nm).

Next shown in figure B.2 is a schematic for a circuit, which allows for ribbon cables to be used to build a magnetic coil. A 20 conductor ribbon cable can then be used to make a 120 turn inductor with only 6 wraps of the ribbon cable. This allows for the magnetic field coil, which provided the quantization axis of the ion to be easily installed and taken off of the experiment.

Finally, a two watt RF source for driving the AOMs is given in figures B.3 thru B.5. Figure B.3 shows the overall circuit diagram of the AOM source. This AOM source is meant to be controlled via analog DC signals (0-10 V). In the experimental setup, this was done by a personal computer with a digital to analog (D/A) output card (see section 4.5). Some of the components of the AOM driver could not be directly sourced, and so they were built and are detailed in the following figures. Figure B.4 shows the schematic for the voltage controlled oscillator circuit, based upon a Crystek VCO module. The VCO circuit is only grounded to the computer's ground and the RF output is coupled by a transformer, so as to avoid ground loops. The various devices in the AOM driver required a number of different analog voltage sources. In figure B.5 is shown a series of DC-DC converters which were used to provide local voltage sources so as to further minimize the possibility of ground loops in the laboratory.

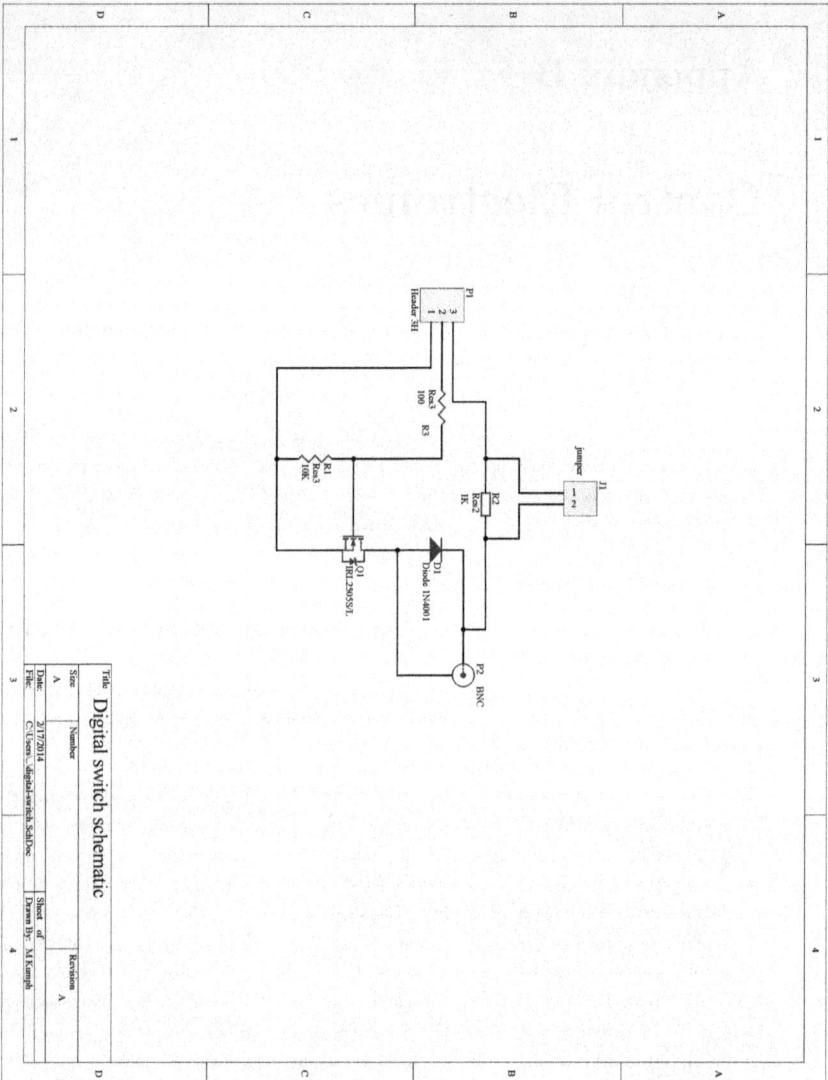

Figure B.1: Digital switch. This digital switch uses a MOSFET to allow the switching of devices with up to 7 A. It was used to control solenoids, which were used as laser shutters and also the calcium oven.

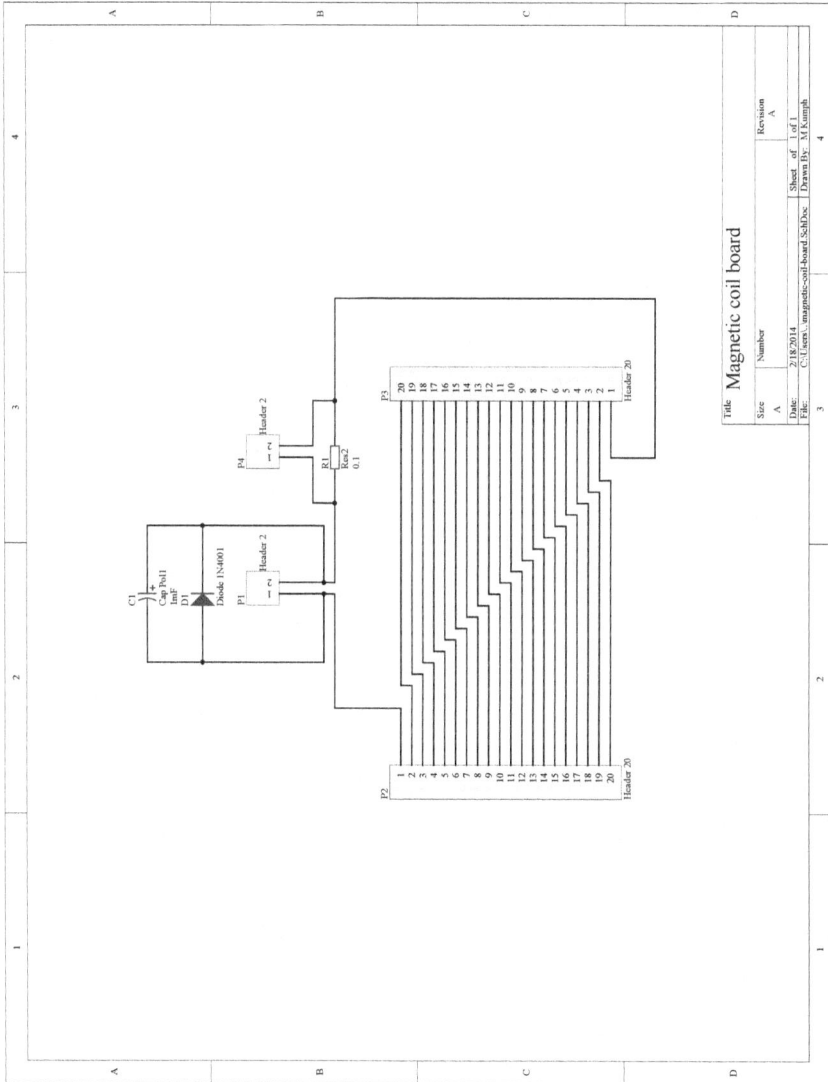

Figure B.2: Magnetic coil board. A printed circuit board was used to simplify the winding of magnetic field coils. By using a 20 pin 0.5 mm pitch ribbon cable, one could wrap the ribbon cable around N times, terminate the ends with connectors and plug it into this circuit board. An inductor with $20 \times N$ turns was then formed.

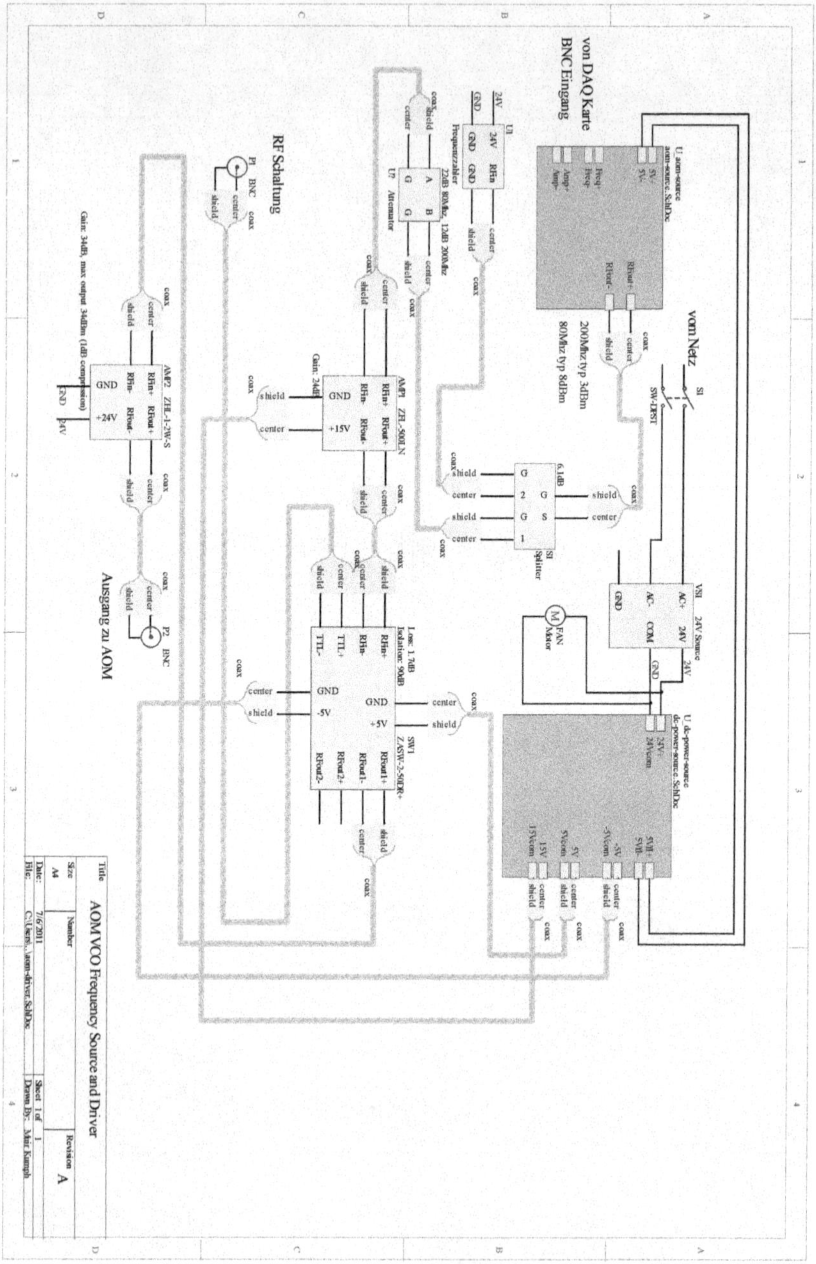

Figure B.3: Acoustic Optical Modulator (AOM) source schematic. A 80/200 MHz 2 W sinusoidal source was built so as to control the frequency and power of the 866, 854, and 397 nm laser beams.

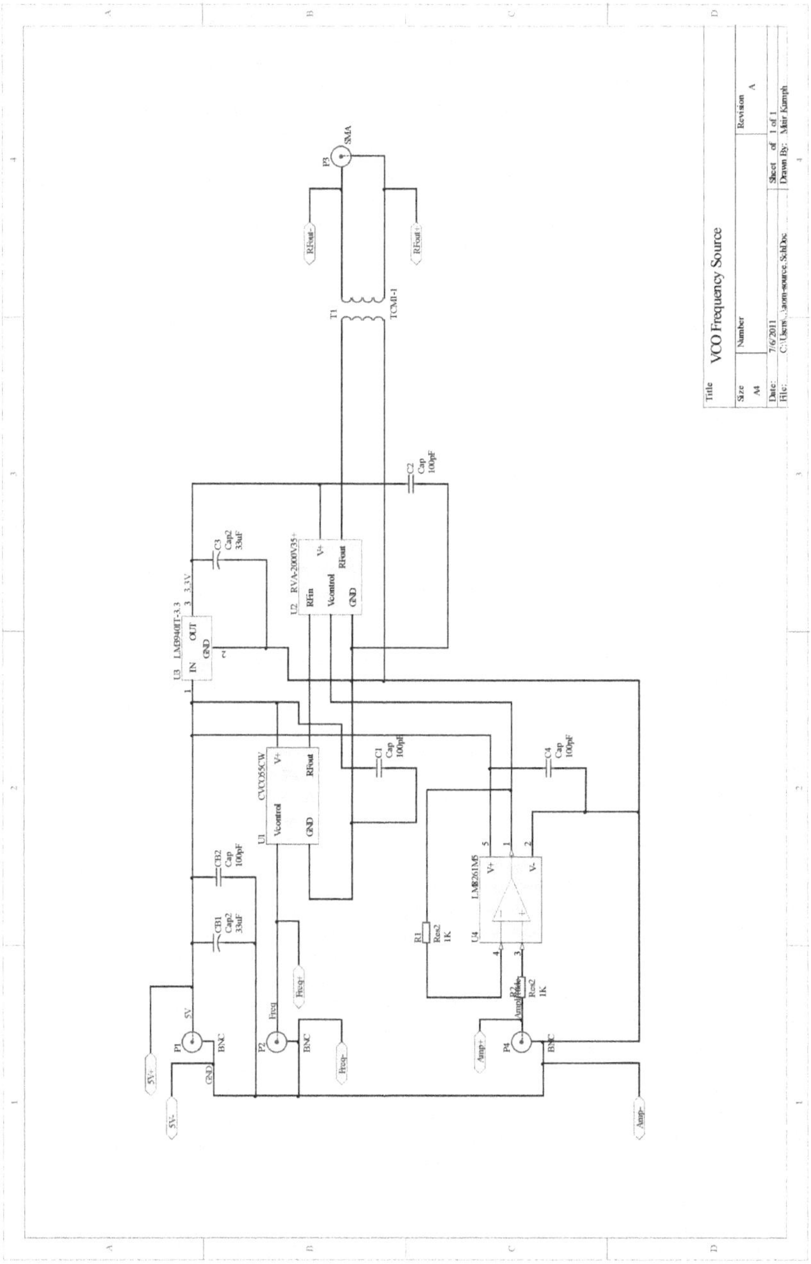

Figure B.4: Voltage Controlled Oscillator (VCO) schematic. The VCO source inside the AOM-source allowed for the frequency and amplitude of the sine waves to be controlled from a digital computer.

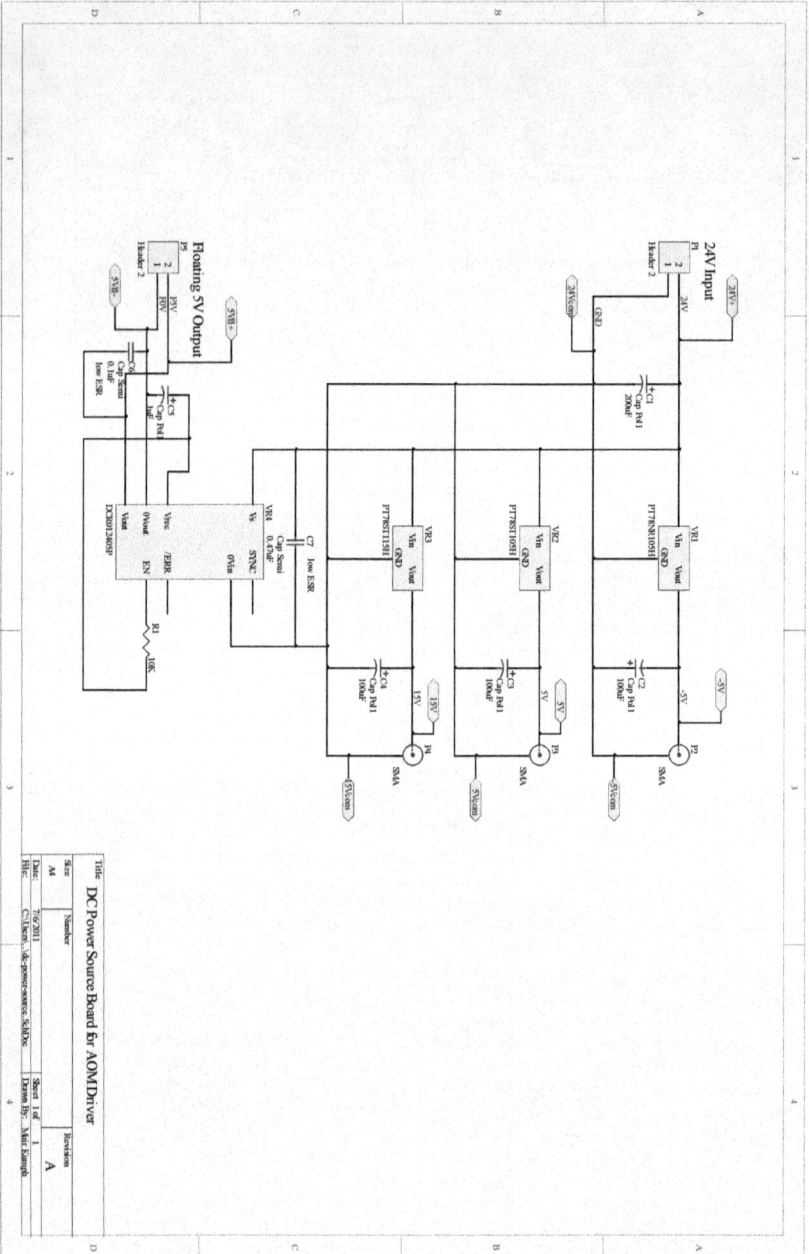

Figure B.5: DC-DC power source schematic. The various components inside the AOM-source required a mix of DC voltages. So that each AOM-source would be DC isolated, each had its own DC power source.

Appendix C

In-Vacuum Electronics

This appendix details the layout of some of the in-vacuum electronics used in the experiments. First in figure C.1 and C.2 is the layout for the break-out printed circuit board (PCB) for the 2D trap array Folsom. Ultra-high-vacuum (UHV) 26 AWG[1] 50 Ω Kapton™ coaxial cables[2] were terminated with UHV coaxial connectors[3]. These cables could then be plugged into the mating connector[4], which were soldered into this PCB. This provided an electrically shielded and strain relieved connection between the trap drive electronics and the 2D array Folsom.

Figure C.3 shows the schematic for the filter board, which Folsom plugged into. Each of the inner 2×2 circular trapping site electrodes was designed with a Π-topology C-R-C filters (cut-off frequency 1.6 MHz) so as to prevent the trap drive from being coupled outside the trap. External 350 kHz 5-pole low-pass filters[5] were also connected to these DC lines to further reduce any noise to the trap.

Figures C.4 and C.5 show the top and bottom copper layers of the PCB based ion trap array Folsom. The trap electrodes are on the top side. Vias (not shown) connecting the electrodes through the PCB allow for a fan-out of the connecting traces on the bottom side. These connecting traces can then terminate to the pin grid array (PGA) pattern of plated through-holes on the outside of the PCB. Each of these plated-through holes allowed a pin[6] to be crimped into the PCB.

Figure C.6 shows the design layout for the microscopic trap array Ziegelstadl v1.0. The design is similar to the Folsom layout. The inner 2×2 array is fully adjustable. The schematic for Ziegelstadl's filter board is shown in figure C.8 and the layout for the filter board is shown in figures C.9 and C.10. It, like in Folsom's design, has a Π-topology C-R-C filter so as to attenuate outside noise on the DC circular trapping site electrodes and prevent RF from being coupled externally.

---

[1] American Wire Gauge
[2] Accu-Glass Products, Inc. part 100720
[3] Allectra Microdot part 245-CON-MIC
[4] Tyco Electronics 1-1532006-3
[5] Mini-Circuits LPF-B0R35+
[6] Millmax 3116

Figure C.7 shows the design layout for the microscopic trap array Ziegelstadl v1.1. The design is similar to v1.0, but it has the following modifications. It has been turned by 4-°, so that lasers cooling beams (from below) can cool in both radial directions of each trap. The trapping site electrode is no longer circular so that the degeneracy of the radial trapping frequencies is also removed. Finally, dedicated bond pads have been drawn away from the vias.

The relatively large capacitance ($\sim 30\,\mathrm{pF}$) of each of the coaxial cables limited the frequency of the RF drive. It was desirable to drive the microscopic 2D arrays of ion traps with a higher frequency, so that another way to wire Ziegelstadl up was developed. Figure C.11 shows a custom vacuum feedthrough for connecting Ziegelstadl. It has a low capacitance shielded path for each separate electrode of the 2D array.

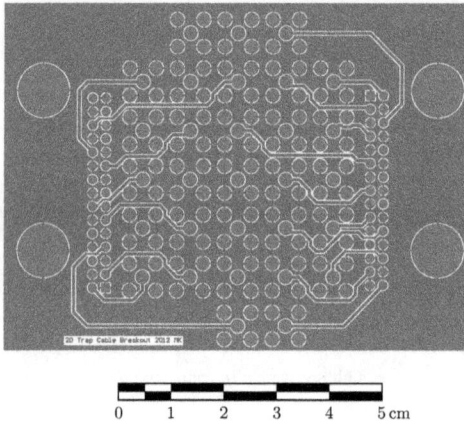

Figure C.1: In-vacuum cable breakout, top layer.

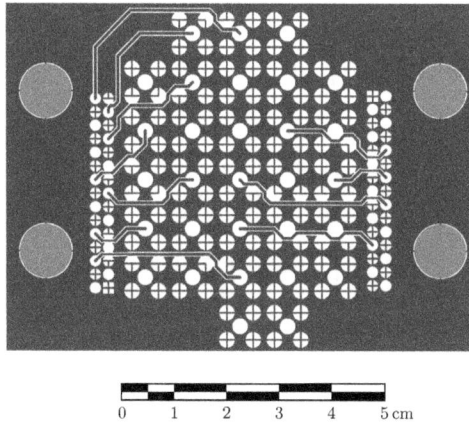

Figure C.2: In-vacuum cable breakout, bottom layer.

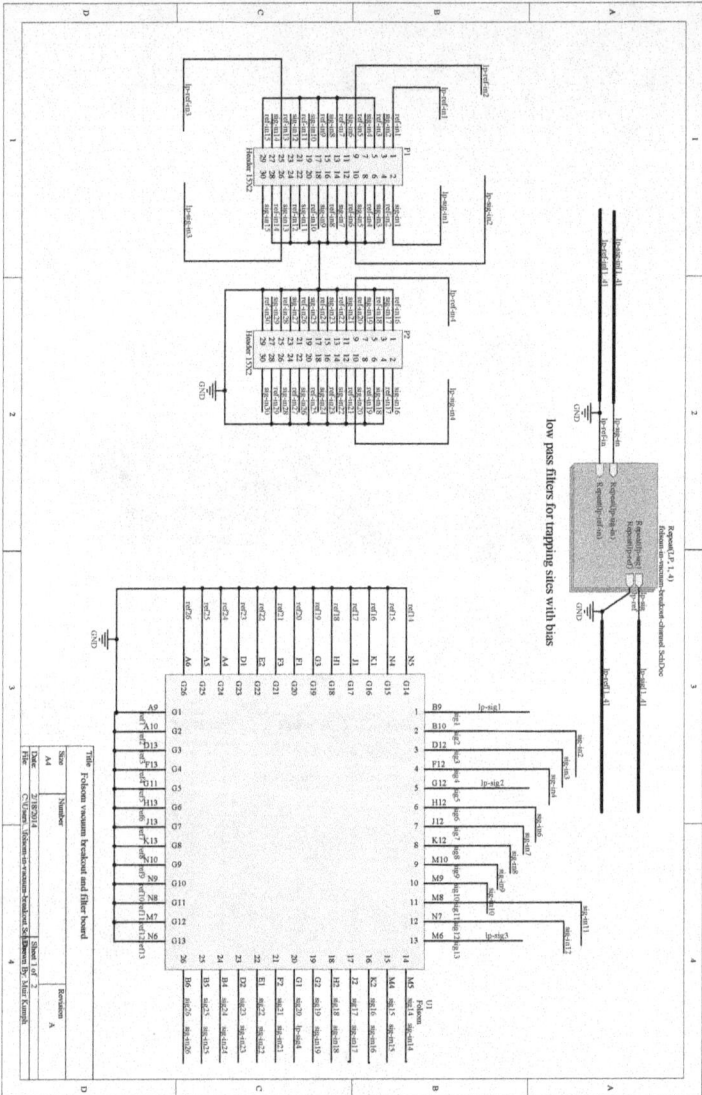

Figure C.3: Folsom filter board. The schematic for the in-vacuum filter board for Folsom includes 4 H-topology C-R-C filters to filter the DC potential of the inner 4 circular trapping sites.

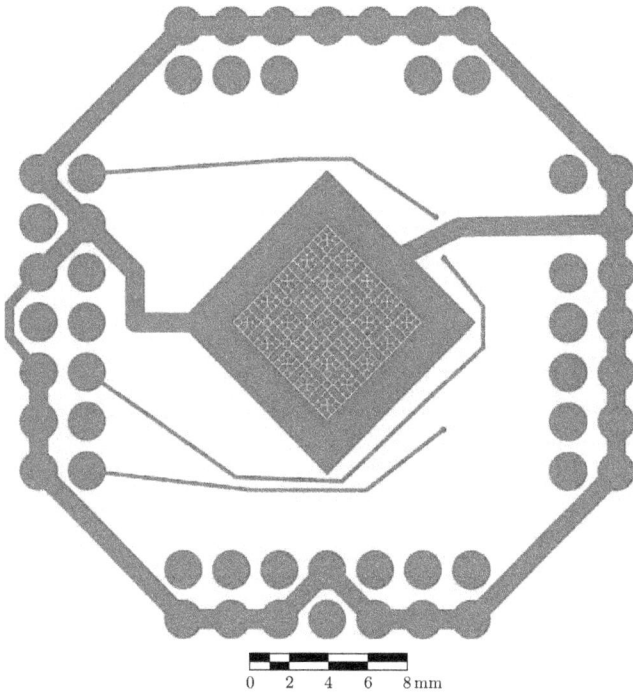

Figure C.4: Folsom trap array, top layer.

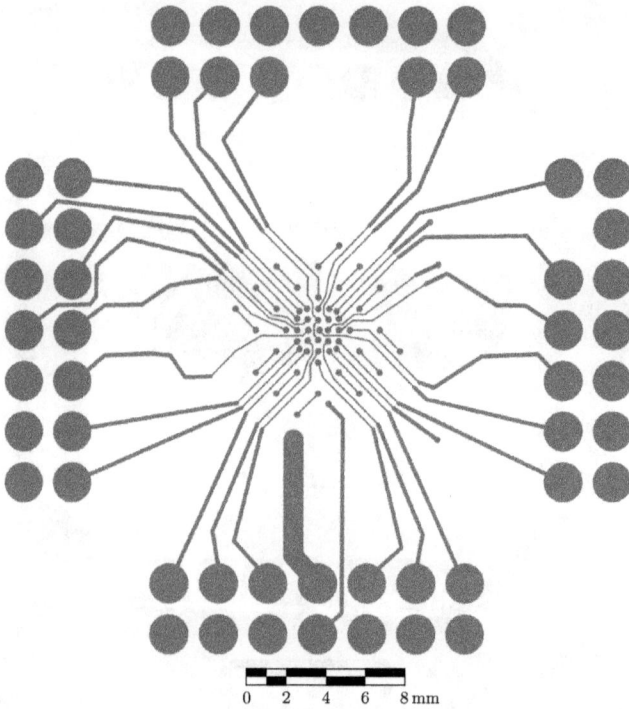

Figure C.5: Folsom trap array, bottom layer.

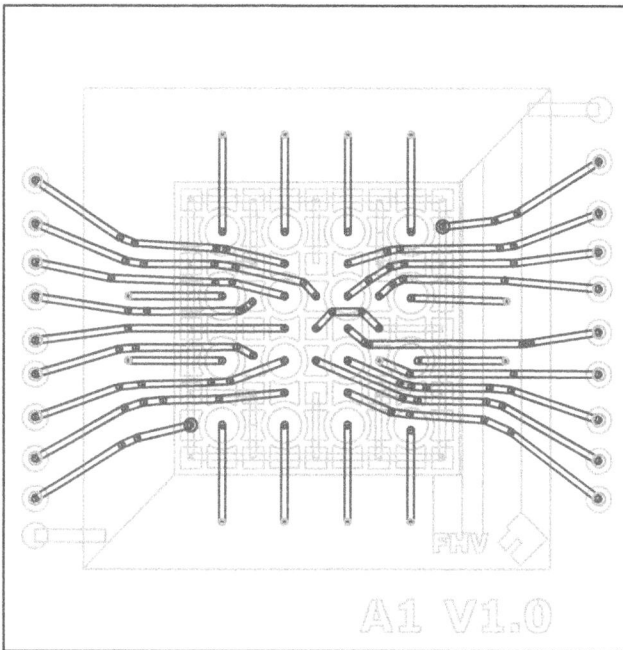

0   50   100 150 200 µm

Figure C.6: Ziegelstadl trap array v1.0, all layers.

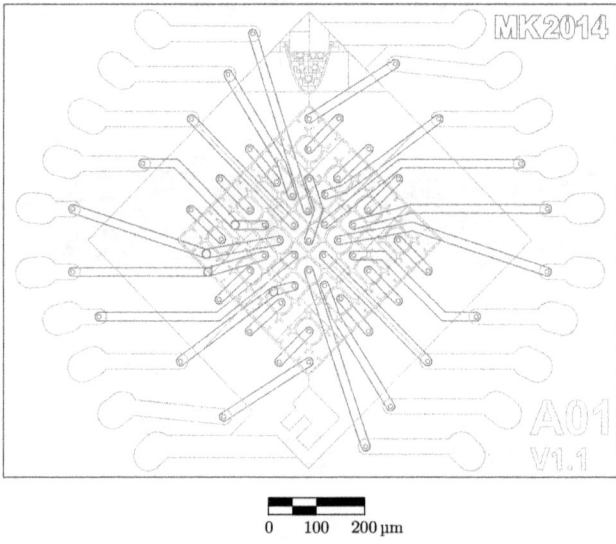

Figure C.7: Ziegelstadl trap array v1.1, all layers.

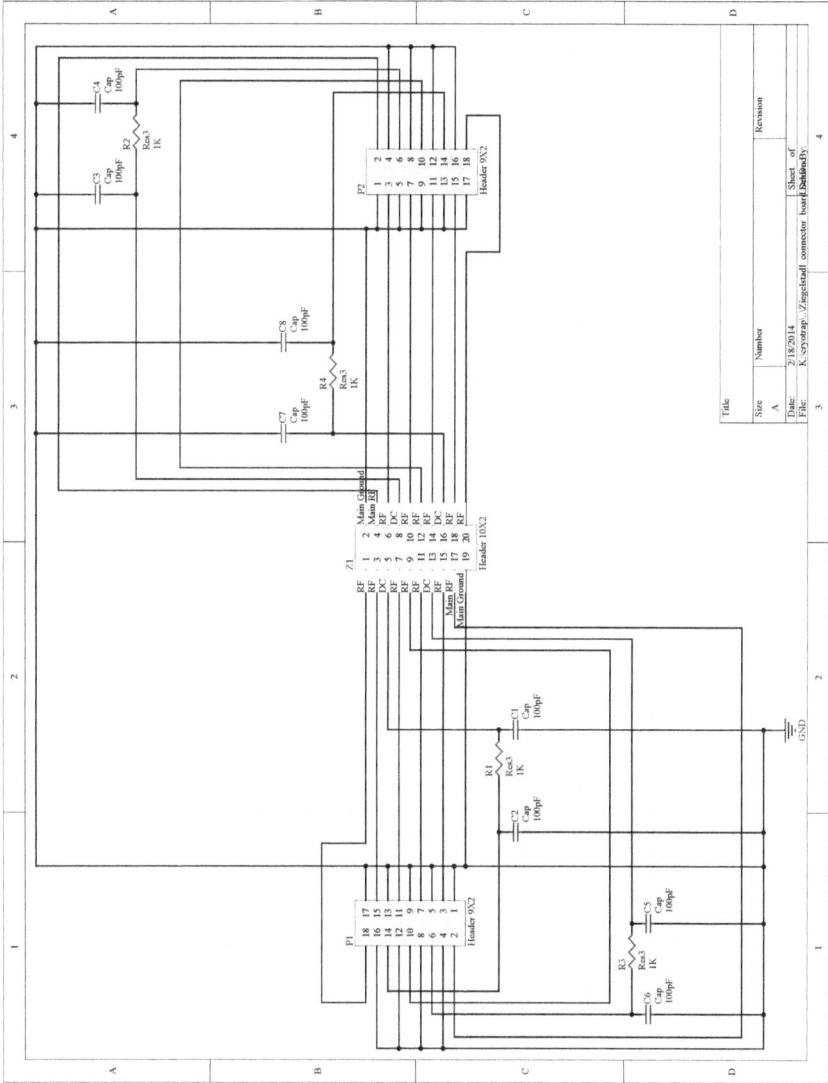

Figure C.8: Ziegelstadl filter board. The schematic for the in-vacuum filter board for Ziegelstadl includes four Π-topology C-R-C filters to filter the DC potential of the inner 4 circular trapping sites.

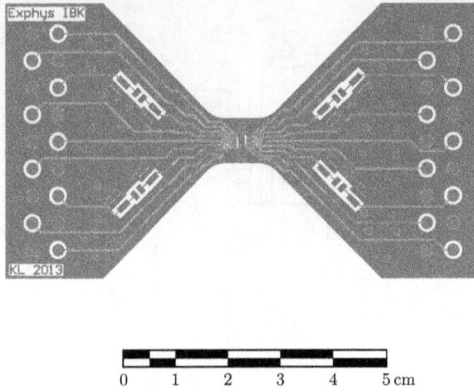

Figure C.9: Filter board and connector board for Ziegelstadl, top side.

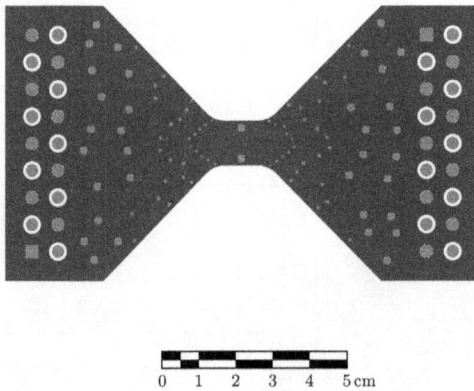

Figure C.10: Filter board and connector board for Ziegelstadl, top side.

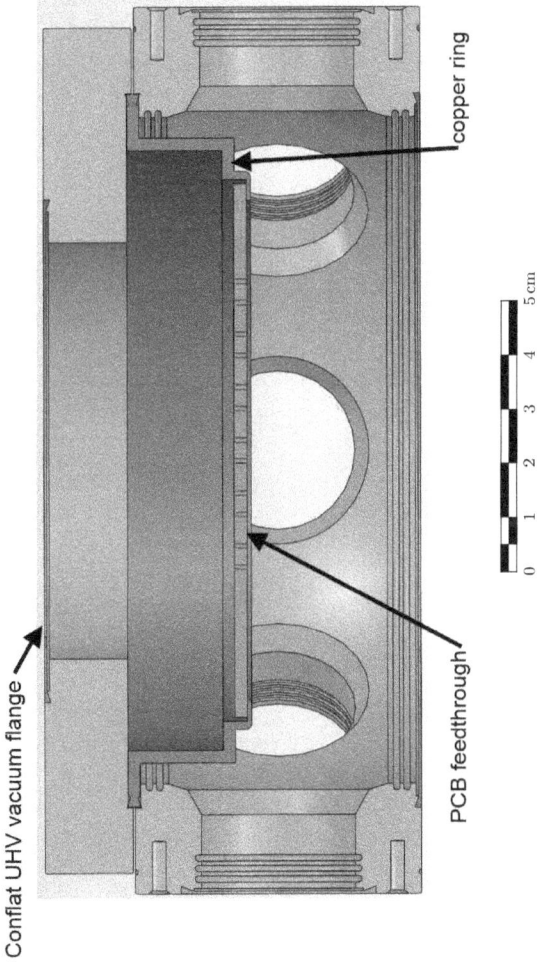

Figure C.11: Low capacitance feedthrough assembly drawing. A printed circuit board was used as a low capacitance feedthrough so as to operate Ziegelstadl. After a bakeout at 130°C, a pressure of approximately 4×10^{-10} mbar was achieved.

Appendix D

Heating Rates in Ion Trap Experiments

In order to understand the sources of heating in ion trap experiments, a thorough review of all known heating rate measurements and their possible causes was undertaken [125]. In this appendix, the list of ion heating rate measurements and the inferred electric-field noise as per equation 2.92 is reproduced.

Figure D.1 shows the spectral density of electric-field noise S_E as a function of the distance d from the ion to the nearest electrode for traps operated nominally at room temperature. Data points are taken from the relevant references in table D.1. On the right, the ordinate scale is given as the equivalent heating rate of a $^{40}Ca^+$ ion with a motional frequency of $\omega_x = 2\pi \times 1\,MHz$. Of importance to the scalability of ion traps, is how the spectral density of the electric-field noise scales as the ion-electrode distance d varies. If one assumes a power law scaling of the form $S_E \propto d^{-\beta}$, then various values of β can be considered. The shaded regions indicate an envelope scaling with d^{-4}; i.e. $\beta = 4$. The dotted lines indicate an envelope scaling with d^{-2}; i.e. $\beta = 2$. See reference [125] for discussion, including the uses and significant limitations of plotting such data on a single graph.

Using this graph purely as a heuristic tool, one can then see there is a trend of increasing electric-field noise as the ion-electrode distance d is reduced. While this graph cannot tell us about any single experiment's source of electric-field noise, in aggregate, the sources of noise which heat ions in these experiments scales as substantially less than d^{-4}. A linear fit to these data points, leads to a distance scaling of $\beta \simeq 2$. While some research has shown a strong increase of the ion's heating rate when the ion-electrode distance is reduced ($\beta = 3.5$ [242]), this collection of all known heating rate measurements offers no evidence that the dominate physical source of noise for the majority of the experiments in this figure scales stronger than d^{-2}. As ion traps are further miniaturized, it will be interesting to see if the scaling trend changes or remains the same.

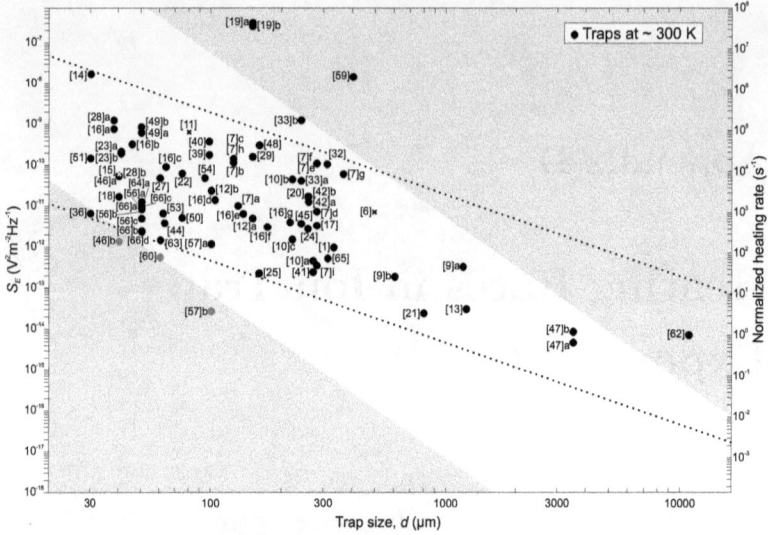

Figure D.1: Spectral density of electric-field noise S_E as a function of the distance d from the ion to the nearest electrode, for traps operated nominally at room temperature. Data points are taken from the relevant references in table D.1. On the right, the ordinate scale is given as the equivalent heating rate of a ^{40}Ca$^+$ ion with a motional frequency of $\omega_x = 2\pi \times 1\,$MHz. The shaded regions indicate an envelope scaling with d^{-4}. The dotted lines indicate an envelope scaling with d^{-2}. See reference [125] for discussion, including the uses and significant limitations of plotting such data on a single graph.

[1]	see [87]	[18]	see [134]	[35]	see [141]	[51]	see [218]
[2]	see [98]	[19]	see [169]	[36]	see [243]	[52]	see [244]
[3]	see [245]	[20]	see [246]	[37]	see [247]	[53]	see [248]
[4]	see [100]	[21]	see [249]	[38]	see [145]	[54]	see [250]
[5]	see [251]	[22]	see [219]	[39]	see [227]	[55]	see [252]
[6]	see [253]	[23]	see [254]	[40]	see [138]	[56]	see [255]
[7]	see [132]	[24]	see [256]	[41]	see [257]	[57]	see [240]
[8]	see [258]	[25]	see [144]	[42]	see [259]	[58]	see [239]
[9]	see [260]	[26]	see [261]	[43]	see [262]	[59]	see [263]
[10]	see [182]	[27]	see [172]	[44]	see [264]	[60]	see [265]
[11]	see [214]	[28]	see [178]	[45]	see [266]	[61]	see [267]
[12]	see [268]	[29]	see [139]	[46]	see [228]	[62]	see [269]
[13]	see [270]	[30]	see [271]	[47]	see [272]	[63]	see [273]
[14]	see [274]	[31]	see [275]	[48]	see [216]	[64]	see [276]
[15]	see [171]	[32]	see [277]	[49]	see [278]	[65]	see [279]
[16]	see [242]	[33]	see [215]	[50]	see [280]	[66]	see [281]
[17]	see [282]	[34]	see [142]				

Table D.1: List of references for trapped-ion heating-rate measurements. The figures and text of this paper denote the traps using the reference numbers here. Where a single publication reports heating rates for several traps, or for one trap under markedly different conditions, these different results are labeled with a letter following the reference number.

Appendix E

Sources of Electric-Field Noise

Electric-field noise is of fundamental importance to quantum information experiments with trapped ions. Because the ion is charged, it reacts to electric fields. electric-field noise can then excite the motion of the ion, which heats it up. The heating equation given in equation 2.92 can be written out for just the first two terms. Considering just the x-direction of an ion confined by a time varying trap drive at frequency $\Omega_{\rm rf}$, the heating rate $\Gamma_{\rm h}$ from the motional ground state of the ion due to incoherent electric-field fluctuations is then given by

$$\Gamma_{\rm h} = \frac{e^2}{4m\hbar\omega_{\rm x}}[S_{\rm E}(\omega_{\rm x}) + \frac{\omega_{\rm x}^2}{2\Omega_{\rm rf}^2}S_{\rm E}(\Omega_{\rm rf} \pm \omega_{\rm x})], \tag{E.1}$$

where $S_{\rm E}(\omega_{\rm x})$ and $S_{\rm E}(\Omega_{\rm rf} \pm \omega_{\rm x})$ are the spectral noise densities of the electric-field noise at the trap frequency $\omega_{\rm x}$ and motional sidebands $\Omega_{\rm rf} \pm \omega_{\rm x}$) respectively. The spectral noise density is defined in equation 2.93. The heating of the ion is then critically dependent upon the electric-field noise at the location of the ion.

As an example, consider an ion with a motional frequency $\omega_{\rm x}$ of $2\pi \times 1\,{\rm MHz}$ and a trap drive frequency $\Omega_{\rm rf}$ of $2\pi \times 30\,{\rm MHz}$ in the presence of an electric-field noise spectral density, $S_{\rm E}$, which is constant over the frequencies of interest. Figure E.1 shows the relative heating rate of the ion, in dB, due to equal intensity spectral noise components at the trap frequency $\omega_{\rm x} = 2\pi \times 1\,{\rm MHz}$ and at the motional sidebands $\Omega_{\rm rf} \pm \omega_{\rm x} = 2\pi \times 30\,{\rm MHz}$. The efficacy of the heating due to noise at the motional sidebands (29 and 31 MHz) is reduced by $\omega_{\rm x}^2/(2\Omega_{\rm rf}^2)$ or about 65 dB compared to the heating from noise at the motional frequency (1 MHz).

This appendix considers various theoretical models describing some of the mechanisms which are thought to contribute to electric-field fluctuations in ion traps. Also considered is how each of these sources of electric-field noise scale with distance, frequency and temperature using the following ansatz

$$S_{\rm E} \propto \omega^{-\alpha}d^{-\beta}T^{+\gamma}, \tag{E.2}$$

where α, β, and γ describe the power scaling dependency of the electric-field noise upon its frequency ω, the ion-electrode distance d, and the trap operating temperature T respectively. Section E.1 considers the noise in free space that would couple directly to an ion in free space,

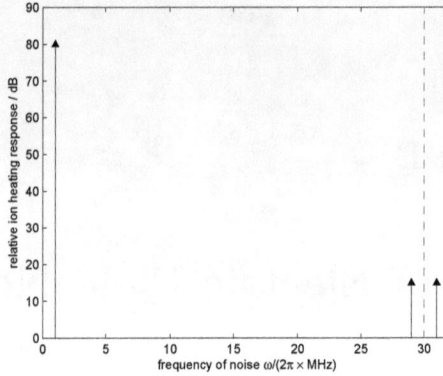

Figure E.1: The relative efficacy of heating, in dB, with a motional frequency ω_x of $2\pi \times 1\,\mathrm{MHz}$ and a trap drive frequency Ω_{rf} of $2\pi \times 30\,\mathrm{MHz}$ in the presence of a spectral noise S_E which is constant over the frequencies of interest. The efficacy of the heating due to noise at the motional sidebands (29 and 31 MHz) is reduced by about 65 dB or a factor of 1800 compared to the heating from noise at the motional frequency (1 MHz).

ignoring any effects from the trap. As well as coupling directly to the ion, such sources of background electromagnetic radiation can couple to wires in the experiment. This can potentially couple noise into the system much more efficiently than by simple free-space coupling directly to the ion. This effect is termed EM pickup and is discussed in section E.2.

Moving on to noise originating in the experimental components themselves, section E.3 considers Johnson-Nyquist noise. In the simplest case, this is the thermal noise of the electrons in resistive elements in the experiment. A number of often-overlooked resistive elements are considered. Additionally, section E.4 considers technical noise, taken here to mean noise from equipment such as power supplies, and coupled to the experiment though the wiring. While this is often not considered strictly Johnson noise, it is included here due to the similarities with distance scaling that it shares with Johnson noise. For EM pickup, Johnson noise, and technical noise the entire surface of the electrode is considered to be at a single (albeit time-varying) potential. Consequently the distance scaling of all three mechanisms is the same and is geometry independent. However, if their are substantial electric-field gradients, such as exist near the RF null of a quadrupole ion trap, the gradient of the field noise can substantially alter the expected distance scaling of the field noise. This is considered in section E.5. A source of noise which can arise during excitation of conductors, called flicker noise, is then considered in section E.6. Next, the long term drift of static potentials is described in section E.7.

The term "patch-potential" refers to a section of a conductor which, for whatever reason, is not at the same potential as the rest of the material. Such a generic term naturally includes a broad range of physical objects, each of which may behave differently depending on the exact details of the patch mechanism. Section E.8 gives an introduction to some possible mechanisms and how they are expected to behave.

Section E.9 considers space-charge, which is when electrical charge starts to fill the space between the electrodes. This is most likely caused by electron emission from the electrode surfaces. And finally in section E.10 the emission of photons at the RF trap drive frequency by the ions is considered. These can cause an exponential rise in the temperature of the ion and are highly dependent on the strength of the ambient electromagnetic field at the RF drive frequency.

E.1   Direct Coupling to Fields in Free Space

A charged particle in a harmonic well is subject to heating by fluctuating electromagnetic (EM) fields and even in the absence of nearby surfaces the ion is affected by freely propagating electromagnetic radiation. The most obvious source for a finite electromagnetic background is black-body radiation. This level is far too low to play a significant role in ion-trap experiments, though it is important to calculate as it provides an important baseline, which is used as a reference for additional noise. Generally, this extra noise above that expected from black-body radiation is termed electromagnetic interference (EMI). Natural sources of EMI are significantly higher than the black-body level, and EMI from manmade sources can be higher still. This section first considers black-body radiation, which provides the baseline, relative to which, all other EMI is given (E.1.1). The levels of EMI expected from both natural and manmade sources are then given (E.1.2). Finally, the scaling of these noise sources with frequency, distance and temperature are discussed (E.1.3).

E.1.1   Black-Body Radiation

If the frequency of the electromagnetic radiation, $\omega \ll k_{\mathrm{B}} T/\hbar$, then the single-sided energy spectral density of black-body electromagnetic radiation in a vacuum is $u_{\mathrm{em}}(\omega) = 2k_{\mathrm{B}} T\omega^2/(\pi c^3)$ [283], where k_{B} is Boltzmann's constant, T is the temperature, and c is the speed of light. [Note: This form of $u_{\mathrm{em}}(\omega)$ has units of energy per unit volume per Hz] The energy density of an electric field, U_{em}, is given by Griffith [284] as,

$$U_{\mathrm{em}} = \frac{1}{2}(\epsilon_0 \mathbf{E}^2 + \frac{\mathbf{B}^2}{\mu_0}) \qquad (\mathrm{E.3})$$

where $\mathbf{E}$ is the electric field, $\mathbf{B}$ is the magnetic field and ϵ_0 and μ_0 are the vacuum permittivity and permeability respectively. Knowing the energy density, U_{em}, one can calculate the electric and magnetic field amplitudes because the electric and magnetic fields contain equal amounts of energy in the far field. Because of orthogonality of sine waves, equation (E.3) is also true for each frequency component of $u_{\mathrm{em}}(\omega)$. Furthermore, the electric-field noise is isotropic, so that each spatial component provides $1/3$ of the total power. We then arrive at the relation

$$\tilde{E}(\omega) = \sqrt{u_{\mathrm{em}}(|\omega|)/(6\epsilon_0)} \qquad (\mathrm{E.4})$$

where $\tilde{E}(\omega)$ is the magnitude of the Fourier transform of a single spatial component of the electric-field amplitude. Since the Fourier transform is defined for positive and negative frequencies, the extra factor of 2 appears in equation (E.4). The definition of the single-sided power spectral density (PSD) (see eq. (2.93)) can be written in terms of the Fourier transform of the signal [285], so that

$$S_E(|\omega|) = 2\tilde{E}^2(\omega). \tag{E.5}$$

We then arrive at the single-sided power spectral density of the black body electric field [286],

$$S_E^{(BB)}(\omega) = \frac{2k_B T \omega^2}{3\epsilon_0 \pi c^3}. \tag{E.6}$$

At room temperature and frequencies of $2\pi \times 1\,\mathrm{MHz}$, this gives a power spectral density of electric-field noise of approximately $10^{-22}\,\mathrm{V}^2/\mathrm{m}^2\mathrm{Hz}$. This is far below the observed level of noise in ion-trap experiments. It is nonetheless important to know this level, as black-body radiation is the reference point for EMI or ambient RF noise, as explained below.

E.1.2 Electromagnetic Interference

Radio-frequency engineers have long known that at low- (30-300 kHz), medium- (0.3-3 MHz) and high- (3-30 MHz) frequency bands, the level of black-body radiation is too low to be responsible for the observed noise level seen on antennas. The ion motional frequency of most ion-trap experiments is in the medium-frequency (MF) band, and such radiation is pervasive: sources as diverse as lightning strikes in the tropics [287], the solar wind hitting the ionosphere [288] and data being sent along power transmission lines [289] cause noise which is near the secular frequency of many ion-trap experiments. Moreover, MF radiation is hardly attenuated by the atmosphere, natural materials, or buildings, and not efficiently radiated into space as it is reflected by the ionosphere.

The standard measure of this EMI is called the external noise factor, F_a, which is a dimensionless factor, defined as the extra noise that a perfect antenna receives above the black-body radiation noise at a temperature of $T = 300\,\mathrm{K}$. The power spectral density (PSD) of the EMI is then $S_E^{(EMI)} = F_a \times S_E^{(BB)}$. Radio engineers have measured the external noise factor's statistics. It varies greatly spatially and temporally. In the MF band, outdoors, away from man-made structures, F_a is about 60 dB [290], and in cities it is approximately 80 dB, decreasing with frequency as ω^{-3}. At 1 MHz, natural sources have an F_a range of 0-100 dB, 99.5% of the time; i.e. 0.5% of the time, natural sources exceed 100 dB. The electric-field noise levels for $F_a = 80$ dB at 1 MHz is then $S_E^{(EMI)} \sim 10^{-14}\,\mathrm{V}^2/\mathrm{m}^2\mathrm{Hz}$, decreasing as ω^{-1}. This level of noise is seen in some ion-trap experiments (cf. Fig. D.1).

Worse yet, inside commercial buildings, where most laboratory experiments would be carried out, there is even more electrical noise at these frequencies than is found outdoors [291]. The external noise factor in commercial buildings has been measured to be approximately 120 dB at 1 MHz and falls off with ω^{-5} [292]. An unshielded ion trap exposed to "typical" levels of indoor noise would therefore see $S_E^{(EMI)} \sim 10^{-10}\,\mathrm{V}^2/\mathrm{m}^2\mathrm{Hz}$ falling off with ω^{-3}.

Of course, trapped ions are surrounded by electrodes, in a vacuum vessel made of metal, which shields them to some extent. Nonetheless, EMI from ambient noise in buildings cannot be ruled out as a source of direct ion heating and care must be taken to shield the ion trap so that

EMI does not affect the experiment. EMI shielding is not a simple engineering chore and EMI can leak in through viewports, poor seams in the shielding and improperly connected grounds. The best way to estimate EMI shielding is through simulation and measurement.

E.1.3 Scaling of Electromagnetic Interference

F_{a} is measured relative to the black body background, where the black-body radiation scales as ω^2 (see eq. (E.6)). This means that the frequency-scaling exponents of F_{a} and $S_{\mathrm{E}}^{(\mathrm{EMI})}$ then differ by 2. For an indoor experiment, where the external noise factor, F_{a}, falls as ω^{-5}, one might expect the spectral density of electric-field noise, $S_{\mathrm{E}}^{(\mathrm{EMI})}(\omega)$, to scale in free space as ω^{-3}. However, the ion is partly shielded by the electrodes, and also shielded by the vacuum chamber and other conductors [such as heat shields in cryogenic systems [170, 293]]. Electric-field shielding via conductors works best at lower frequencies (the opposite being true for magnetic-field shielding) and so, while shielding can provide substantial reductions in the expected heating rate, it would also change the expected frequency scaling of the electric-field noise. The effect of this EMI shielding should be taken into account. Furthermore, any strong sources at specific frequencies in the noise spectrum have not been included in this analysis: radio stations and electronic devices can emit strong signals which would be further above the levels mentioned above. These would manifest themselves as peaks in the power spectrum and would be ephemeral and/or geography dependent.

The heating due to direct electromagnetic interference would be expected to be independent of the trap size provided the shielding did not change as a function of trap geometry. In practice, changing the geometry of the electrodes will almost certainly change the shielding at the ion, unless care is taken to shield the entire experiment from EMI. Some non-zero value of the distance-scaling exponent, β would therefore be expected, but this would have to be calculated for each specific apparatus.

The background black-body radiation level scales linearly with temperature (see eq. (E.6)). However, the level of environmental EMI impinging on an experiment would not be expected to vary with temperature. That said, if the ion trap has metallic shielding which is cooled, the shielding's effectiveness would change as its conductance increases with decreasing temperature. At low temperatures this will eventually reach a plateau when the conductance no longer changes.

E.2 Electromagnetic Pickup

As explained in section E.1, above, electric fields from EMI can directly interact with an ion in a trap. As well as interacting with the ion directly, it is also possible that EMI causes voltage fluctuations on the wires and electronics of the experiment, which then carry this noise to the trap. These voltage fluctuations caused by EMI coupling to wires are referred to as electromagnetic (EM) pickup, and are considered in detail here. The electric component of EMI can be well shielded by metallic conductors. However, low-frequency magnetic-field noise from EMI is not as easily shielded and can be picked up by any conductive loops in the trap wiring. An example of where such loops may occur in ion traps is the wiring leading to the RF electrodes: in a linear-trap configuration, a single RF source is usually split to drive two trap electrodes. The capacitance between the two RF electrodes then completes the loop. According to Faraday's law, any changing magnetic flux perpendicular to the surface formed by a loop will induce an electro-motive force (EMF) around the loop. Since the small capacitance between the two trap electrodes has a very large impedance, this is where the full voltage drop of the induced EMF will present itself.

The frequency scaling, α, of noise due to pickup of electromagnetic radiation (EM pickup) would be dependent upon the scaling of the underlying EMI radiation. Beyond this, it must be taken into account that the voltage V_L, at frequency ω, induced in a loop of wire from an incident magnetic field, B, normal to the loop, also at frequency ω, is given by Kanda et al. [294] as,

$$V_{\mathrm{L}} = \omega B A_{\mathrm{L}} \tag{E.7}$$

where A_L is the loop area. The spectral density of the magnetic-field noise S_B can be calculated using equation (E.3) and using the Fourier transform definition of the power spectral density in equation (E.5). The EM pickup voltage noise $S_V^{(\mathrm{PU})}$ is then,

$$S_{\mathrm{V}}^{(\mathrm{PU})} = F_{\mathrm{a}} \frac{2\mu_0 k_{\mathrm{B}} T A_{\mathrm{L}}^2 \omega^4}{3\pi c^3}. \tag{E.8}$$

This means that, for the same free-field noise, α for EM pickup would be smaller by two than for direct coupling of the free fields to the ion (see eq. (E.6)). Thus, if the EMI scaled with $\alpha = 3$ (as is expected for noise indoors) then the EM pickup would scale with $\alpha = 1$.

At a given frequency and temperature, the noise on the electrodes caused by EM pickup is indistinguishable – from the ion's point of view – from Johnson noise. Consequently, the spectral density of electric-field noise above the surface scales as D^{-2} where D is the characteristic length scale of the electrode. This is discussed more fully under Johnson noise in section E.3.

As EM pickup is caused by the EMI interaction with the electronics of the ion trap, temperature might, at first, not seem to be relevant. However, if there is any shielding in the experiment, the shielding's effectiveness is dependent upon temperature. Typically, as temperature is lowered, the resistance is lowered and the shield becomes more effective at damping magnetic fluctuations, so that for a typical experiment $\gamma > 0$.

The above discussion of EM pickup explains what sort of power spectral density of voltage noise can be expected on a trap electrode. The electric-field noise, to which this will give rise at the position of the ion, is explained below under Johnson noise in section E.3. For completeness, we give an estimate here of the absolute noise level expected from EM pickup: Considering an environment (typical office building) with $F_a = 120\,\mathrm{dB}$, the voltage noise due to pickup on an unshielded 10 cm diameter loop of wire will be $S_V^{(\mathrm{PU})} \sim 10^{-18}\,\mathrm{V}^2/\mathrm{Hz}$. This is equivalent to the Johnson noise of a $200\,\Omega$ resistor at room temperature. If this voltage noise is on a trap electrode, with a characteristic length scale of $D = 100\,\mu\mathrm{m}$, then the electric-field noise will be $S_E^{(\mathrm{PU})} \sim 10^{-10}\,\mathrm{V}^2/\mathrm{m}^2\mathrm{Hz}$. This level will scale in proportion to the square of the area enclosed by the loop.

E.3 Johnson-Nyquist Noise

Johnson-Nyquist noise (or Johnson noise) is the electrical noise generated by thermal motion of charge carriers in a conductor. It has been shown [295, 296] that the power spectral density of such voltage noise is given by

$$S_{\mathrm{V}}^{(\mathrm{JN})}(\omega) = 4k_{\mathrm{B}}TR(\omega, T), \qquad (\mathrm{E}.9)$$

where $R(\omega, T)$ is the effective real resistance, at frequency ω, of the whole circuit from the two terminals across which the voltage noise is observed. What constitutes the *whole circuit* is broadly a matter of convention. Some analyses consider only the trap electrodes, others consider the associated passive electronics, and still others the full system up to and including active components such as power supplies. The Johnson noise due to the bulk resistance of the trap electrodes themselves is almost always negligibly low. However, as shown below, Johnson noise can appear due to losses in other elements in the experiment that are often overlooked.

Some authors lump many noise sources together, modelling both technical noise (see sec. E.4) and EM Pickup (see sec. E.2) as Johnson noise. This can be done by considering equation (E.9) and replacing the resistance, R by a (much) larger effective resistance [297], or alternatively, a resistance at a (much) higher effective temperature [202]. Here, we shall consider Johnson noise only from electrical elements on the circuit which leads to the trap electrodes.

The spectral density of the electric-field noise due to Johnson-Nyquist noise is

$$S_{\mathrm{E}}^{(\mathrm{JN})} = \frac{S_{\mathrm{V}}^{(\mathrm{JN})}}{D^2} = \frac{4k_{\mathrm{B}}TR(\omega, T)}{D^2}, \qquad (\mathrm{E}.10)$$

where D is the characteristic length scale of a particular electrode in the ion trap.

A simplistic interpretation of Johnson noise may assume that it exhibits an essentially flat noise spectrum ($\alpha = 0$) and that the D^2 term in the denominator of equation (E.10) would imply $\beta = 2$. Given many metals have $R \propto T$, one might also anticipate $\gamma = 2$. Finally, assuming trap electrodes to have a resistance of $\sim 0.1\,\Omega$ one might expect the electric-field noise due to Johnson noise at $300\,\mathrm{K}$, in a trap of $D = 100\,\mu\mathrm{m}$ to be around $S_{\mathrm{E}}^{(\mathrm{JN})} \sim 10^{-13}\,\mathrm{V}^2/\mathrm{m}^2\mathrm{Hz}$. Such an analysis is, however, rather too simplistic to describe the situation in realistic ion-trap experiments.

In what follows we consider a number of physical effects which cause the Johnson-noise analysis to be more complex than the simple picture just presented. First, the frequency dependence of Johnson noise is considered (see sec. E.3.1). We then consider the distance scaling (sec. E.3.3) - specifically how the characteristic length, D, relates to the ion-electrode separation, d, for various geometries. This is followed by sections on the temperature scaling (E.3.4) and absolute levels (E.3.5) of Johnson noise. Finally, consideration will be given to technical noise (see sec. E.4).

E.3.1 Frequency-Dependent Resistance

It can immediately be seen from equation (E.10) that the frequency scaling, α, of Johnson noise would depend on the frequency scaling of $R(\omega)$ which, for standard ideal resistors, is flat i.e. $\alpha = 0$. However, electrical circuits connected to the trap almost always contain frequency-dependent impedance elements such as filters. Consequently, such frequency dependencies must be considered carefully as it is, in principle, possible to achieve any value of α.

Since we are considering ion heating due to the electric-field noise at the trap's secular frequency, the Johnson-noise induced heating would be dependent upon the real part of the impedance seen from the trap electrodes at that frequency [190]. It is easy to underestimate this resistance and it is nearly always much higher than the direct current (DC) resistance of the circuit. This is because frequency-dependent elements are always lossy; even superconductors have an AC resistance [298].

Idealized capacitors, inductors and transmission lines (wires) have, as their main characteristics, only reactances (imaginary impedances). However, the materials which make up these devices are not perfect and have losses. The impedance of a wirewound inductor is an obvious example: a wirewound inductor typically has a very low DC resistance, but if the current is kept constant while the frequency of the current is increased, the inductor will heat up.

Real dielectrics are also lossy, characterized by a loss tangent, $\tan\theta$. Any capacitance incorporating a dielectric has an equivalent parallel resistance (EPR) given by Vandamme et al. [299] as,

$$\text{EPR} = \frac{1}{\tan(\theta)\omega C} \tag{E.11}$$

where ω is the angular frequency and C is the capacitance of the capacitor. A large EPR would be considered a high quality capacitor and have little associated noise. Only by modelling the rest of the circuit connected to the capacitor, can the equivalent series resistance (ESR) be computed. The ESR of the whole circuit is then easily understood as the resistance which goes into equation (E.10) to determine the Johnson noise.

While capacitors used in the trap electronics may be the most obvious instance of such loss in dielectrics, the dielectrics which hold the trap must also be considered. Macroscopic traps often have their electrodes held by Macor [260, 300] which has been shown to cause significant dissipation in ion-trap systems [205]. Surface traps usually consist of thin ($\sim 1\,\mu$m) metallic electrodes on dielectric substrates such as quartz [171], silica [227] or sapphire [215]. Excellent (low-loss) dielectric materials have loss tangents of $\sim 10^{-5}$. Typically the ion-trap electrodes have a capacitance of just a few picofarads, which means that at trap frequencies of $2\pi \times 1\,$MHz, even in the best case, the equivalent parallel resistance will not be higher than $10\,$GΩ. If such a trap electrode were to be isolated from its voltage source at the trap frequency, this would give rise to an ESR of $\simeq 1\,\Omega$. To keep Johnson noise below such a level requires careful attention to the drive electronics and filters so that they absorb this noise from the electrodes, while filtering out the noise invariably coming from the power supplies.

Losses from magnetic components can be treated in a very similar manner (again, characterized by a loss tangent). This can be seen, for instance, in the case of a layer of nickel, which is used as a barrier layer between gold and copper. Being magnetic it can cause substantial losses at microwave frequencies [301]. This is less likely to be a significant issue for ion traps as, in most ion-trap systems, magnetic materials are generally avoided due to concerns over magnetic-field instability affecting the ion's electronic states. Most commonly, magnetic components in ion trap systems are wire-wound inductors inside filters or resonators which lead to the trap electrodes. The quality factor Q relates an inductor's real and imaginary impedances as

$$Q = \frac{\omega L}{R_{\mathrm{L}}}, \tag{E.12}$$

where L is the inductance of the inductor and R_{L} is the real part of the inductor's impedance. The loss tangent being related to the quality factor by $\tan \theta = 1/Q$. If a filter inductor, $L = 100\,\mu\mathrm{H}$, is meant to work at, $\omega = 2\pi \times 1\,\mathrm{MHz}$, has a quality factor, $Q = 20$, then its effective resistance is $\sim 30\,\Omega$ at this frequency. This resistance would then need to be correctly accounted for in any Johnson noise analysis.

Eddy-current losses in conductors near a wire carrying RF and microwave frequencies can also cause the effective resistance of the wire to increase substantially above that expected from the bulk DC resistance [302]. At medium RF frequencies, this effect would be only on the order of $\simeq 1\,\mathrm{m}\Omega$ per cm of wire, a distance $\simeq 100\,\mu\mathrm{m}$ away from a non-magnetic conductor. This effective resistance could be enhanced by several factors. It increases roughly with the logarithm of the magnetic permeability of the nearby conductor. Multiple turns of the current carrying wire would increase the effect roughly with the logarithm of the number of turns squared. At $1\,\mathrm{MHz}$ the effect saturates at around $0.1\,\Omega/\mathrm{cm}$. However, as the frequency is increased, the effective resistance due to eddy-current losses scales proportionally without limit. Details are given in [302]. A situation in ion trap experiments where this effect should be taken into account, is if a shield is placed over a helical resonator or wire-wound inductor. The resulting drop in the inductor's Q is due to these eddy current losses.

E.3.2 Johnson Noise Near the Trap Drive Frequency

If the noise on the RF electrodes near the trap drive frequency Ω_{rf} is to be calculated, care must be taken when computing the effective resistance of RF electrodes to ground. This is because the presence of the RF resonator can particularly enhance the effective resistance of the RF electrodes to ground. For instance, if a trap is operated with a drive frequency of $29\,\mathrm{MHz}$ and a motional frequency of $2\,\mathrm{MHz}$, has a capacitance of $5\,\mathrm{pF}$ and with a loss tangent of 0.01 and is made with a resonator that has an unloaded Q of 200, the effective Johnson noise resistance can of order $1\,\mathrm{k}\Omega$ at the motional sidebands $\Omega_{\mathrm{rf}} \pm \omega_{\mathrm{x}}$. This would give rise to a voltage noise S_{V} at the motional sidebands with a level of approximately $10^{-17}\,\mathrm{V}^2/\mathrm{Hz}$. Because the RF electrodes produce a quadrupole field with an RF null at the expected position of the ion, their voltage fluctuations are not usually considered. However, as explained below (sec. 2.11.2), if the ion is displaced from the RF null by a stray field, it will experience excess micromotion and their will be heating due to common voltage fluctuations on the RF electrodes.

E.3.3 Characteristic Length Scale

The characteristic length of a system, D, determines the conversion of voltage to electric-field fluctuations as defined in equation (E.10). Electric-field noise measurements in ion traps usually report the minimum ion-electrode distance, d. While this is often much simpler to determine, it is generally quite different from the characteristic length. Indeed, for some noise sources and geometrical symmetries the characteristic length can tend to infinity. For instance the RF electrodes create a symmetric field near the RF null. So that if the common-mode voltage fluctuations on the RF electrodes are considered, the characteristic length scale of the RF electrodes D_{RF} is always infinity at the RF null. However, if the ion is displaced from the RF null, then D_{RF} is not infinity and depends on both the ion's displacement and the gradient of the quadrupole field.

The electric field due to voltages on electrodes in vacuum can be computed using the Laplace equation

$$\nabla^2 \phi = 0, \tag{E.13}$$

where ϕ represents the potential everywhere in space and the voltages on the electrodes represent the boundary conditions for solving this differential equation. Then the electric field is then given by $\mathbf{E} = \nabla \phi$.

In general, there can be several characteristic length scales, corresponding to the different components of the electric field and which electrode in a trap has voltage noise on it. If a voltage is applied to the j-th electrode, and the resulting spatial electric-field component i is calculated, then the characteristic length scale is

$$D_{i,j} = V_j / E_{i,j}, \tag{E.14}$$

where $E_{i,j}$ describes the i-th component of the electric field due to a voltage on the j-th electrode. Closely related to the characteristic dimension is the dimensionless quantity, the dipole efficiency factor, $\kappa_{i,j}^{(\mathrm{D})}$, which relates the ion-electrode separation to the characteristic distance.

$$\kappa_{i,j}^{(\mathrm{D})} = \frac{2d_j}{D_{i,j}} = \frac{2d_j E_{i,j}}{V_j} \tag{E.15}$$

where d_j is the distance of the ion from the j-th electrode. The extra factor of 2 in the above equation ensures that the efficiency is equal to 1 for a pair of parallel plates, where the ion is trapped half-way between.

For the simple geometries considered in this section, symmetry provides that there is only one direction that the ion can be heated. This being due entirely to just one voltage (or voltage difference) of interest. For the rest of this section, no subscripts are needed for $\kappa^{(\mathrm{D})}$, d, D, V, or E. If the geometry of a trap were to be uniformly scaled, then $\kappa^{(\mathrm{D})}$ is constant, $d \propto D$, and $\beta = 2$. If, however, some dimensions of a trap are changed while others remain constant, then one must solve for the electric field as a function of the changing geometry to calculate $\kappa^{(\mathrm{D})}$, D and β. There exist a limited number of geometries for which the characteristic distances can be analytically calculated; in general it must be found numerically. Three geometries are considered here: surface, spherical and needle geometries. In the first instance they are calculated analytically. A more complicated – and realistic – needle geometry is then calculated numerically.

E.3.3.1 Planar Geometry

An often-used idealization for determining the effects of voltage noise on the electrodes is to approximate the trap by two parallel and infinitely extended plates [68, 77, 202, 286]. This configuration can approximate the situation for an ion trapped above a planar microtrap, or more generally describes the limit in which the ion-surface distance is much smaller than any extended features of the trap electrode or the curvature of the electrode. The electric field normal to two plates separated by a distance L, due a voltage difference V, is simply V/L. In any real geometry consisting of large but finite planes there exists some far-field ground, such as the vacuum chamber. As the theoretical model considers infinite planes there is no quadrupole term and no pseudopotential minimum in which the ion might be trapped. The position of the ion between the plates is thus arbitrary, though we can choose to place the ion at $d = L/2$, so that the characteristic length scale is $D = L = 2d$ and $\kappa^{(D)} = 1$. The distance scaling of the electric-field noise is then $\beta = 2$.

E.3.3.2 Spherical Geometry

Another idealized geometry is to compose the trap of two conducting spheres, a distance L apart, of equal radius, r_0, and a far-field ground. In this geometry the ion is trapped halfway between the two spheres. Provided $L \gg r_0$, the charge distribution on the surface of the spheres will give the same electric field half way between the two spheres as that of point charges at the spheres' centres. This configuration is sometimes used to approximate a needle quadrupole ion trap (QIT) or the end caps of a linear QIT. The electric-field strength at the center of the ion trap would then be,

$$E = \frac{CV}{4\pi\epsilon_0 d^2} \tag{E.16}$$

where C is the capacitance of the two conducting spheres, V is the voltage difference between them, and $d = L/2$. The capacitance of one sphere in such a two sphere system, $C \simeq 2\pi\epsilon_0(r_0 + r_0^2/L)$ [303], allows for an analytical solution of the field to be found in the limit $L \gg r_0$. The characteristic length scale is then $D = 2d^2/r_0$, the dipole efficiency factor is $\kappa^{(D)} = R/d$, and the distance-scaling exponent is $\beta = 4$.

E.3.3.3 Needle Geometry

Another geometry which has been studied in ion-trap experiments is that of two needle electrodes [242]. The ion is trapped halfway between the two needle tips. The distance, d, between the ion and the needle tip can be varied, and the expected electric-field noise due to voltage noise on the electrodes can be calculated by solving the Laplace equation. If the needle electrodes are approximated by hyperbolic surfaces of revolution (see Fig. E.2a), then the Laplace equation for the electric field can be solved in prolate spheroidal coordinates [304]. This approximation has the advantage that it can be solved analytically, though it suffers from the limitation that the radius of curvature of the needle tips cannot be independently specified from the taper angle of the needle tips. Similarly to the planar geometry above, the absence of a far-off ground in the mathematical model means that no ion trap is formed with the application of a common-mode RF voltage to the tips. However, for the purposes of calculating the heating rate due to voltage noise on the needle tips, it does provide some insight.

Considering two hyperbolic needle tips with radius of curvature r_0, separated by a distance $2d$, the electric field half-way between them is given by,

$$E = \frac{V}{d} \frac{v_0}{\ln(\frac{1+v_0}{1-v_0})} \tag{E.17}$$

where $v_0 = \sqrt{1/(1 + r_0/d)}$, and V is the potential difference between the two tips. This simplified needle geometry has a characteristic length scale $D = d \ln(\frac{1+v_0}{1-v_0})/v_0$, which can be fitted to a power law, in order to estimate the distance scaling, β, for a particular trap geometry. Approximating equation (E.17) over the range $30 \, \mu m < d < 200 \, \mu m$ with $r_0 = 3 \, \mu m$ with a power law $d^{-\beta}$ gives $\beta \cong 2.4$. This is sensitively dependent on parameters such as tip radius of curvature r_0 and the exact range of distance d. For the above example, if d has an uncertainty of $\pm 5 \, \mu m$ and r_0 has an uncertainty of $\pm 1 \, \mu m$, then $\beta = 2.4(2)$.

Another method of calculating the electric field due to a voltage difference between the two tips is to use finite-element modelling (FEM) techniques. A number of commercial software programs can perform these calculations, which allow for a more realistic electrode geometry to be simulated. Using COMSOL[1] to simulate a geometry similar to that described by Deslauriers et al. [242] gives results shown in Fig. E.2. The figure shows the results for an analytical solution to the electric field between two needle tips, as well as two-dimensional (2D) and (three-dimensional) 3D FEM simulations. The simulated geometry (see Fig. E.2b) has needle electrodes with a tip radius of curvature $r_0 = 3 \, \mu m$ and a taper angle of 4°. Also included are ground sleeves of 3 mm inside diameter, recessed 2.3 mm from the needle tips and electrically isolated from the needles.

As can be seen from Fig. E.2, the different methods give significantly different results. The difference between the analytical and the FEM solutions can be explained by the difference in the geometry calculated: the analytical solution was unable to include the effects of the ground sleeves. Because of the rotational symmetry, a cylindric 2D simulation should suffice. The difference between the two- and three-dimensional FEM calculations is attributable to the coarse-grained mesh used in 3D, set by the limits of reasonable computer memory. As the 3D mesh was refined, the results tended ever closer towards the 2D result. Taking the 2D simulation result, then, as the most reliable, it can be seen that the dipole efficiency factor, $\kappa^{(D)}$, changes considerably, by as much as a factor of two, over the range plotted. Consequently, $d \not\propto D$

---

[1]COMSOL Multiphysics v3.4

(a) Analytical geometry

(b) FEM geometry

(c)

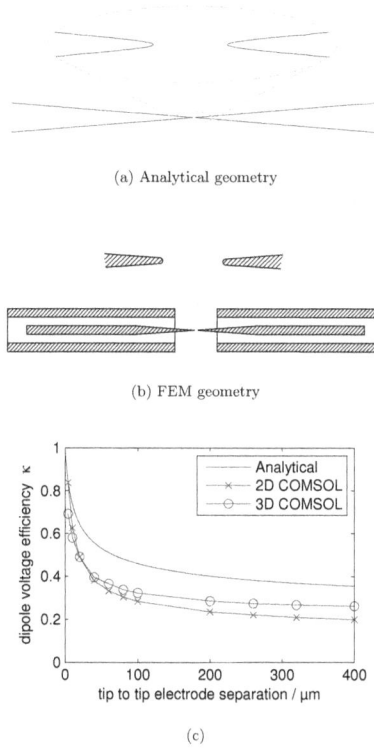

Figure E.2: Simulations of a needle trap. (a) and (b) show the geometry used to simulate the results of calculating the electric field at the center of a needle trap, where the needle tips have a radius of $3\,\mu$m. For the two- and three-dimensional COMSOL simulations, the taper angle was $4°$ and a conducting ground sleeve, $3\,$mm inside diameter, was recessed $2.3\,$mm from the tips. This geometry was chosen to approximate the experiment of Deslauriers[242]. (c) results for the analytical and numerical simulations of the trap.

and $\beta \neq 2$. The distance-scaling exponent can then be calculated by fitting the characteristic distance, D, to a power law. Assuming that the measured ion-electrode separation has an uncertainty of $\pm 5\,\mu$m [which is the uncertainty claimed by Deslauriers et al. [242]] then the predicted value of β is 2.5(2).

E.3.4 Temperature Scaling of Johnson Noise

From equation (E.10) it should be expected that the level of Johnson noise varies with temperature. This picture is complicated by the temperature dependence of the various resistances involved in the system. Many metals' resistance varies as $R \propto T$. This would suggest, in the simplest case, that $\gamma = 2$. That being said, at low temperatures, materials can depart significantly from such simple behaviour [305]. This applies both to the simple materials such as the trap electrodes and also to any electronic components which may be held at cryogenic temperatures. Finally, it should be noted that even if the trap itself is cooled to very low temperatures, not all of the attendant electronics may be at the same temperature. This will complicate the picture of any temperature scaling.

Ultimately, each individual experiment must be modelled properly to check what value of γ would be expected, and how that value might change in different temperature regimes.

E.3.5 Absolute Level of Johnson Noise

The absolute level of electric-field noise (and associated heating) that could be expected from an effective $2\,\Omega$ resistance of an electrode to ground is considered here. This level of effective resistance is achievable with careful attention to electronics. Furthermore it is assumed that the trap has a trap drive frequency $\Omega_{\rm rf} = 2\pi \times 30\,{\rm MHz}$, and is operated at a secular frequency of $\omega_{\rm x} = 2\pi \times 1\,{\rm MHz}$. For an effective resistance of $2\,\Omega$ at the electrode, the voltage noise $S_{\rm V}$ is approximately $10^{-20}\,{\rm V^2/Hz}$. Since Johnson noise, for a fixed resistance, has a flat spectral density, this level of voltage noise would be seen at both the trap frequency $\omega_{\rm x}$ and the micromotion sidebands $\Omega_{\rm rf} \pm \omega_{\rm x}$. However, since heating is suppressed at the micromotion sidebands by a factor of $\omega_{\rm x}^2/(2\Omega_{\rm rf}^2)$ (see eq. E.1), the Johnson noise is not usually considered at the micromotion sidebands.

Considering for now, just noise at the motional frequency $\omega_{\rm x}$, a trap with a characteristic distance D of $1\,{\rm mm}$ and $2\,\Omega$ of resistance would then give rise to an electric-field noise level of approximately $10^{-14}\,{\rm V^2/m^2Hz}$. This level would correspond to a heating rate of approximately 1 phonon/sec for $^{40}{\rm Ca}^+$. This would not be a concern for most experiments.

E.4 Technical Noise

Technical noise is defined here as noise coming from power supplies and other voltage sources such as digital-to-analog (DAC) cards in experiments. It arises from the imperfect nature of the power supplies in laboratory equipment. While the physical mechanism of the technical noise from power supplies is usually Johnson-Nyquist or EM pickup in nature and amplified up by electronics, it is considered separately here. Technical noise can alternatively be modeled as a resistor which is very hot [202] or very large [297].

Frequency scaling, α, of technical noise could be anything as the device could exhibit resonances. However, many DC power supplies and DACs used for experiments have specifications, which allow an experimenter to put an upper limit on any such technical noise. A typical power supply might have $5\,{\rm mV}$ of noise spread across $20\,{\rm MHz}$ ($1\,\mu{\rm V/Hz^{1/2}}$), which is, using equation (E.9) equivalent to the Johnson noise on a $75\,{\rm M}\Omega$ resistor. One would need to use roughly $80\,{\rm dB}$ of filtering at the trap frequency to lower the electric-field noise so that this technical noise becomes roughly equal to the Johnson noise expected from a bulk resistance of $\sim 1\,\Omega$. With this level of filtering and a characteristic distance $D = 1\,{\rm ,mm}$, a spectral field noise of $10^{-14}\,{\rm V^2/m^2Hz}$

could then be expected at the position of the ion. Such aggressive filtering is possible, but not trivial, and it is easy for such filter electronics to have low Q inductors (see eq. (E.12)) or loops of wire subject to EM pickup (see sec. E.2) which would then have Johnson noise, which is larger than the filtered technical noise.

Distance scaling, β, of technical noise, would be $1/D^2$ since it is proportional to the voltage noise on the electrodes. The temperature scaling, γ, should be flat, but again if the filters change their response as the temperature changes, one would expect a non-zero γ.

E.4.1   Spectral Purity of the RF Drive

Related to the technical noise in DC voltage sources is the noise in function generators. Signal generators, which are used to drive the RF electrodes of ion traps, produce noise in addition to the desired trap drive signal. The noise can be decomposed into phase noise and amplitude noise [306]. The frequency of the desired sine wave is called the carrier at frequency Ω_{rf}. Phase noise is the impurity of the phase of the desired signal that a signal generator produces. The amplitude of the carrier can also have fluctuations called amplitude noise. Phase and amplitude noise are often expressed in the units dBc, which is decibels relative to carrier [307]. The sum of both the amplitude noise and the phase noise gives the total generator's noise.

The primary source of concern for the RF drive signal sources is that ions can be heated by electric-field noise at the first order motional sidebands at frequencies of $\Omega_{\mathrm{rf}} \pm \omega_{\mathrm{x}}$. If the RF electrodes are driven separately, then each could create field noise at the position of the ion, corresponding to its characteristic distance D. The heating rate at the motional sidebands is reduced by a factor of $\omega_{\mathrm{x}}^2/(2\Omega_{\mathrm{rf}}^2)$ compared to the same level of noise directly at the motional frequency ω_{x} (see eq. E.1). As an example, consider an ion trap with a motional frequency $\omega_{\mathrm{x}} = 2\pi \times 1\,\mathrm{MHz}$ and an RF electrode with a characteristic distance D of 1 mm being driven from a sine wave source at a drive frequency $\Omega_{\mathrm{rf}} = 2\pi \times 30\,\mathrm{MHz}$ so that, for the same intensity of electric-field noise, the heating response at the motional sidebands $\Omega_{\mathrm{rf}} \pm \omega_{\mathrm{x}}$ is about 65 dB less than at the motional frequency ω_{x} (see fig. E.1).

One method to estimate the maximum spectral purity of a signal generator is by considering the Johnson noise from a $50\,\Omega$ source and then adding any amplifier gain factor and noise factor (NF). For instance, the noise floor of a $50\,\Omega$ source is -174 dBm at a temperature of 300 K. There is usually some amplifier and electronics which might amplify this noise by 44 dB. For the trap parameters just given above, this would give an electric-field noise of approximately $10^{-11}\,\mathrm{V^2/m^2Hz}$ at the 1 MHz motional sidebands ($\Omega_{\mathrm{rf}} \pm \omega_{\mathrm{x}}$). For the example given here the attenuation of the heating at the motional sidebands $\omega_{\mathrm{x}}^2/(2\Omega_{\mathrm{rf}}^2)$ is approximately 1/1000, compared to the same level at the motional frequency ω_{x}. This level of noise at the micromotion sidebands would be equivalent to $10^{-14}\,\mathrm{V^2/m^2Hz}$ at the trap's secular frequency ω_{x} and would correspond to a heating rate of 1 phonon/sec. Such a level of heating, would not likely be of concern to quantum information experiments. Unfortunately, the signal generators used to drive RF electrodes have more technical noise than that expected from above. This extra technical noise combined with stray static fields and the attendant micromotion, can form a significant source of heating.

Typical modern DDS-based function generators have non-harmonic phase noise of about -130 dBc 1 MHz away from the carrier [308]. The amplitude noise has the same magnitude for random fluctuations of the RF. This means that a trap with a 1 MHz secular frequency, being driven directly by a 100 V_{RMS} signal source, would have to contend with approximately 10^{-9} V^2/Hz of spectral voltage noise at the first order motional side-bands ($\Omega_{rf} \pm \omega_x$) known to cause heating. Fortunately, ion traps nearly universally use an RF resonator to match the signal generator's source impedance to the trap's (mostly capacitive) load (see section 6.2 for a discussion). This resonator also acts as a band-pass filter and depending on the quality factor of the resonator, the trap's capacitance C, and how the matching network is designed, one might obtain an additional 40-50 dB of attenuation 1 MHz away from the carrier (Q=500, C=20 pF). So that one would expect 10^{-14} V^2/Hz of noise 1 MHz away from the carrier at the RF drive electrodes. Continuing with the above example, if there were just a single RF electrode with a characteristic distance D of 1 mm, this would give a field noise of approximately 10^{-8} V^2/m^2Hz at frequencies of $\Omega_{rf} \pm \omega_x$. For a trap, with a trap drive frequency Ω_{rf} to secular frequency ω_x ratio of 30, this would give a heating rate equivalent to an electric-field noise at the secular trap frequency of $S_E(\omega_x)$ of 10^{-11} V^2/m^2Hz. This would give rise to a noticeable amount of heating for experiments of about 1 phonons/ms (assuming ^{40}Ca$^+$). For ion traps driven with multiple RF electrodes, each with its own signal generator, heating of this type could be a concern.

It would seem that ion traps, driven with a single RF signal source for all RF electrodes would not be subject to heating due to technical noise on the motional sidebands, as the RF field (and its field noise) are both zero at the trap location. However, if the ion is displaced from the RF null, it will experience field noise due to the gradient of the RF field noise. Below, the effect of a field noise gradient is considered.

E.5 Field Noise Gradient

The above analysis assumed that the field noise S_E can be considered spatially homogeneous near the ion's location. This is not true if one is concerned with the field noise due to common-mode voltage fluctuations on the RF quadrupole electrodes. If a single signal source drives all of the RF electrodes which form a quadrupole potential at the center of the trap, the field noise is zero at the center of the trap, but it grows with the distance from the RF null x and the electric field's gradient α_t. If the ion oscillates about the RF null, symmetry dictates that only the even order secular and micromotion heating terms need be considered.

Consider the heating in the x direction, where the gradient of an RF quadrupole field α_t is constant. The root-mean-square electric field is then $E_x^{rms} = \alpha_t x_D$, where x_D is the RMS displacement of the ion from the RF null. The gradient of the electric field α_t can be written in terms of trap geometry and applied voltage as,

$$\alpha_t = \frac{V_{RF}}{D_Q^2}, \tag{E.18}$$

where V_{RF} is the RMS voltage on the RF electrodes and D_Q^2 describes the quadrupole characteristic distance of the trap geometry. The power spectral density of the noise at the location of the ion S_E will then be

$$S_E = S_V \frac{x_{\mathrm{D}}^2}{D_{\mathrm{Q}}^4}. \tag{E.19}$$

Considering just the lowest even order heating terms, the second order secular heating is due to field noise at $2\omega_{\mathrm{x}}$ and the second order motional sideband heating is due to field noise at $\Omega_{\mathrm{rf}} \pm 2\omega_{\mathrm{x}}$. Such heating caused by the squeezing of the trapping potential is called parametric heating and is detailed in Savard et al. [131]. This type of second order heating, while growing exponentially, is negligible for well cooled ions. However, if a stray field causes the ion to be pushed away from the RF null, then the ion will see a non-zero level of field noise due to voltage fluctuations on electrodes which form the quadrupole potential, which is described next.

E.5.1 Excess Micromotion and Motional Sidebands

As described above, because of the symmetry of the quadrupole fields produced by the RF electrodes, there would be no first order motional sideband heating at $\Omega_{\mathrm{rf}} \pm \omega_{\mathrm{x}}$ due to noise on the RF drive when the ion oscillates about the RF null (see section E.3.3). However, if the ion is displaced from the RF null by a static field, the field noise must be reconsidered. For instance, consider a stray field that pushes the ion away from the RF null by $x_{\mathrm{D}} = 1\mu\mathrm{m}$ in a trap with a quadrupole characteristic distance of $D_{\mathrm{Q}} = 100\,\mu\mathrm{m}$. The electric-field noise at the ion can then be computed using equation E.19. For the case discussed above in this section where the voltage noise on the RF electrodes is $10^{-14}\,\mathrm{V}^2/\mathrm{Hz}$, the electric-field noise would then be approximately $10^{-10}\,\mathrm{V}^2/\mathrm{m}^2\mathrm{Hz}$ at frequencies of $\Omega_{\mathrm{rf}} \pm \omega_{\mathrm{x}}$. Furthermore, because the ion is displaced from the RF null in a field-noise gradient, the excess micromotion of the ion mixes down the field noise at the frequency of the first order motional sidebands $\Omega_{\mathrm{rf}} \pm \omega_{\mathrm{x}}$. In the ion's reference frame, it then experiences additional field noise at the motional frequency ω_{x} [145, 202, 309] with a level scaled by the factor $\omega_{\mathrm{x}}/(2\Omega_{\mathrm{rf}}^2)$. This mixed-down excitation adds *in-phase* with the noise at the first order motional sidebands ($\Omega_{\mathrm{rf}} \pm \omega_{\mathrm{x}}$), so that the heating is a factor of four higher than that expected from just the level of field noise at the motional sidebands $S_E(\Omega_{\mathrm{rf}} \pm \omega_{\mathrm{x}})$ alone [125]. For ion's with excess micromotion, the first order motional term in equation 2.92 should then be modified to include the mixed-down noise from the field-noise gradient. So that, in the presence of this gradient, the heating rate of an ion $\Gamma_{\mathrm{h}}^{(\mathrm{grad})}$ is then given as,

$$\Gamma_{\mathrm{h}}^{(\mathrm{grad})} = \frac{e^2}{4mh\omega_{\mathrm{x}}} \left[S_{\mathrm{E}}^{(\mathrm{grad})}(\omega_{\mathrm{x}}) + 4\frac{\omega_{\mathrm{x}}^2}{2\Omega_{\mathrm{rf}}^2} S_{\mathrm{E}}^{(\mathrm{grad})}(\Omega_{\mathrm{rf}} \pm \omega_{\mathrm{x}}) \right], \tag{E.20}$$

where $S_{\mathrm{E}}^{(\mathrm{grad})}$ is the spectral density of the electric-field noise at center of the ion's motion, due to the displacement of the ion away from the RF null.

Continuing with the example above, where there is a level of voltage noise on the RF electrodes of approximately $10^{-14}\,\mathrm{V}^2/\mathrm{Hz}$. If one assumed that heating was only due to electric-field noise at the trap's secular frequency, then it would seem as if there was excessive field noise at the secular trap frequency at a level of approximately $10^{-13}\,\mathrm{V}^2/\mathrm{m}^2\mathrm{Hz}$. Moreover, if the trap's characteristic distance is changed, while keeping the motional frequency and any uncompensated stray fields constant, the field noise and the associated heating would scale roughly as $1/D_{\mathrm{Q}}^4$.

This has the same predicted scaling as patch potentials (see below in sec. E.8). The combination of spectral impurity in the RF source and uncompensated static stray fields can cause excessive heating of the ion that has a distance scaling of D^{-4}, which is different than that expected from Johnson or technical noise (D^{-2}).

E.6 Flicker Noise

Flicker noise (also called *1/f* noise) is a fluctuation phenomena, where the power spectral density of the fluctuations $S^{(\text{flicker})}(\omega)$ of a physical quantity falls-off as the frequency ω increases. The range of the fall-off is quite broad despite the name $1/f$. Flicker noise falls off as $S^{(\text{flicker})}(\omega) \propto 1/\omega^{\alpha}$, where $0 < \alpha < 2$ [310]. It is found in many physical systems, but was first documented in electrical systems by Johnson [311].

E.6.1 Electronic $1/f$ Noise

The physical mechanisms of electronic flicker noise remain, to this day, unknown. This is despite nearly a hundred years of research into the physical origins of this noise. Many years of experiments have ruled out some models. And despite the fact that it was first observed in a system typically associated with surface science, flicker noise is often considered a bulk effect [312, 313].

Narrowing the discussion to flicker noise in metal conductors, this phenomenon can be modelled as a time varying change in the bulk resistance of a conductor, $R(t)$, at time t [313, 314]. The single sided power spectral density (PSD) of the resistance fluctuations S_{R} of a conductor is defined as

$$S_{\text{R}}(\omega) = 2 \int_{-\infty}^{\infty} d\tau \langle \delta R(\tau) \delta R(0) \rangle e^{-i\omega\tau}, \tag{E.21}$$

where $\delta R(t) = R(t) - \bar{R}$ is the variation in the conductor's resistance from its average value $\bar{R}$.

The PSD of the resistance of a wide variety of metal conductors $S_{\text{R}}^{(\text{flicker})}(\omega)$ has been measured to fall-off roughly as $1/\omega^{\alpha}$, where $0.9 < \alpha < 1.5$. A DC current excitation of the metal conductor is most often used to observe and measure it. Many of the features of flicker noise across a resistor due to a DC excitation can be described by the phenomenological equation due to Hooge [312]

$$S_{\text{V}}^{(\text{flicker})}(\omega) = \frac{\bar{V}_{\text{DC}}^2 \gamma_{\text{H}}}{[\omega/(2\pi)]N}, \tag{E.22}$$

where S_{V} is the observed spectral density of voltage noise (given in V^2/Hz) due to an average DC voltage $\bar{V}_{\text{DC}}$ across the conductor, N is the number of charge carriers in the sample, and γ_{H} is the material-dependent dimensionless Hooge parameter. Measurements of many metals indicate that the value of the Hooge parameter is roughly 2×10^{-3}. Equation E.22 can also be rewritten using Ohm's law to express the power spectral density of the resistance fluctuations as

$$S_R^{(\text{flicker})}(\omega) = \frac{R^2 \gamma_H}{[\omega/(2\pi)]N},$$ (E.23)

where $S_R^{(\text{flicker})}(\omega)$ is the spectral density of the resistance fluctuations of the conductor (given in Ω^2/Hz) with an average resistance $\bar{R}$.

In the literature, usually a DC current is used to measure flicker noise. And since ion traps would not normally have any DC current flowing through their electrodes, it might seem as though this source of noise would not need to be considered. However, systems with time varying current, at a frequency Ω_{rf}, exhibit two kinds of noise that are associated with the flicker noise observed with a DC excitation. The first is noise, which is seen around the carrier, when the conductor is excited with a time varying current at carrier frequency Ω_{rf}. Such noise around the carrier is called $1/\Delta f$ noise. The second is a low frequency $1/f$ noise, which manifests itself although there is no DC current [315].

E.6.2 $1/\Delta f$ Noise

Consider a conductor with a time varying current at a frequency of Ω_{rf}. The electrical noise, at a frequency of $\Omega_{\text{rf}} \pm \delta\omega$, due to flicker noise of the conductor is called $1/\Delta f$ noise [316]. The corresponding level of $1/\Delta f$ noise around the carrier frequency Ω_{rf} can be predicted by the $1/f$ resistance fluctuations (eq. E.23), where the frequency ω is replaced with the distance from the carrier frequency $\delta\omega = \omega - \Omega_{\text{rf}}$,

$$S_R^{1/\Delta f}(\Omega_{\text{rf}} \pm f) = \frac{\bar{R}^2 \gamma_H}{[\delta\omega/(2\pi)]N}.$$ (E.24)

For ion traps, the RF drive can cause $1/\Delta f$ noise around the drive frequency Ω_{rf}. Since flicker noise is inversely proportional to the number of charge carriers N, it can be pronounced in thin film conductors, or in high frequency circuits where the current is confined due to the skin effect.

As an example, consider an ion trap, with thin-film metal RF electrodes (for instance 200 μm x 5 mm x 100 nm), with 2 pF of capacitance to ground, and a resistance of 2 Ω, being driven at 30 MHz with 100 V_{rms}. This would create an RMS current I_{RMS} of ~ 38 mA through the electrode. Figure E.3 shows the model used for estimating the flicker noise in the thin film electrode. The trap is driven with a resonant circuit (where the resonator's inductance cancels the trap's capacitance), so that variations in the trap electrode's resistance cause a corresponding change in the RMS current flowing in the circuit. Here, the resistance of the thin-film metal electrode is assumed to be the predominant loss in the system, so that the RF resonator and signal source are matched to drive this 2 Ω load. This means that at steady state, when $\omega_x \ll \Omega_{\text{rf}}$, the RMS trap drive current variation due to a change in the resistance of the thin-film electrode is 1/2 the change of the resistance; i.e. a 2% change in resistance from flicker noise, would result in a 1% change in the RMS current. Using this relationship, one can then relate the $1/\Delta f$

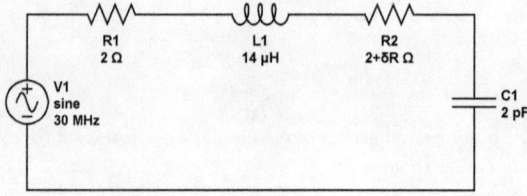

Figure E.3: The schematic used for estimating the flicker noise in the thin film electrode. The trap is driven with a resonant circuit (where the resonator's inductance cancels the trap's capacitance), so that variations in the trap electrode's resistance δR cause a corresponding change in the RMS current flowing in the circuit. Here, the resistance of the thin-film metal electrode is assumed to be the predominant loss in the system, so that the RF resonator and signal source are matched to drive this $2\,\Omega$ load. This means that at steady state, when $\omega_x \ll \Omega_{rf}$, the RMS trap drive current variation due to a change in the resistance of the thin-film electrode is $1/2$ the change of the resistance; i.e. a 2% change in resistance from flicker noise, would result in a 1% change in the RMS current.

current noise to the RMS current I_{RMS} flowing through the thin film resistor as

$$S_I^{(\text{flicker})} = \frac{I_{RMS}^2 \gamma_H}{4[\delta\omega/(2\pi)]N}. \tag{E.25}$$

The voltage noise on the trap is then the capacitive reactance of the trap electrodes at the trap drive frequency $X_C = -1/(\Omega_{rf}C)$ ($\sim 2.6\,\text{k}\Omega$) squared times the current noise $S_I^{(\text{flicker})}$.

Figure E.4 shows the relative heating response of the ion overlayed with the expected $1/\Delta f$ and $1/f$ voltage noise of this driven thin film conductor. The $1/\Delta f$ noise is centered around the carrier frequency $\Omega_{rf} = 2\pi \times 30\,\text{MHz}$. The motional sidebands at $\Omega_{rf} \pm \omega_x$ can be excited by noise at these frequencies (29 and 31 MHz).

Table E.1 gives the expected levels of $1/\Delta f$ noise for various metals at the motional sidebands $\Omega_{rf} \pm \omega_x$, with the above geometry and operating conditions. From the table, the flicker noise on the trap electrodes at the 1 MHz motional sidebands is substantial ($\sim 10^{-15}\,\text{V}^2/\text{Hz}$) when compared to the expected Johnson noise from a $2\,\Omega$ resistor (merely $\sim 10^{-20}\,\text{V}^2/\text{Hz}$). If the RF electrode is driven individually, like in the 2D arrays described in this thesis, then the electric-field noise at the ion will then be of order

$$S_E^{(\text{flicker})} \simeq \frac{S_V^{(\text{flicker})}}{D^2}, \tag{E.26}$$

where $S_V^{(\text{flicker})}$ is the voltage noise on the electrode and D is the electrode's characteristic distance. If the ion is in a trap, with a characteristic distance D of about 1 mm, then an electric-field noise level of approximately $10^{-9}\,\text{V}^2/\text{m}^2\text{Hz}$ at the motional sidebands $\Omega_{rf} \pm \omega_x$ is to be expected at the position of the ion due to $1/\Delta f$ noise. The efficacy of heating at the motional sidebands is reduced by a factor of $\omega_x^2/(2\Omega_{rf}^2)$, when compared to noise at ω_x (see eq. E.1). This level of electric-field noise would then produce a motional heating of approximately 100 phonons/sec for $^{40}\text{Ca}^+$. This would be a source of concern for high fidelity motional gates.

Even for systems where the RF electrodes are driven together, flicker noise could be a source of concern. One likely possibility is that some substantial fraction of the RF current

material	γ_{H}	N $\times 10^{10}$	$S_{\mathrm{I}}(\Omega_{\mathrm{rf}} \pm \omega_{\mathrm{x}})$ A^2/Hz	$S_{\mathrm{V}}^{(\mathrm{flicker})}(\Omega_{\mathrm{rf}} \pm \omega_{\mathrm{x}})$ V^2/Hz
Cu	3.8×10^{-3}	1.1	1.2×10^{-22}	0.9×10^{-15}
Ag	3.5×10^{-3}	0.7	1.8×10^{-22}	1.3×10^{-15}
Au	1.2×10^{-3}	0.9	0.5×10^{-22}	0.3×10^{-15}

Table E.1: The expected flicker noise at 1 MHz from the RF trap drive carrier frequency given for an electrode with a capacitance of approximately 2 pF to ground and a series resistance of $2\,\Omega$, being driven at 30 MHz with $100\,\mathrm{V_{rms}}$. The electrode geometry is a thin rectangular film of dimensions $200\,\mu\mathrm{m}$ x $5\,\mathrm{mm}$ x $100\,\mathrm{nm}$. The number of free charge carriers was computed using the Hall effect constants of the materials.

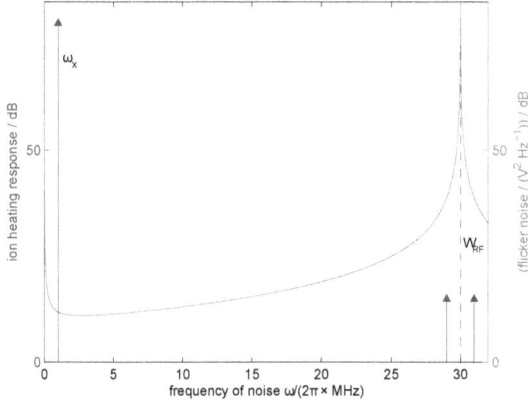

Figure E.4: The relative heating response of the ion overlayed with the expected $1/\Delta f$ and $1/f$ voltage noise of a driven thin film conductor (described in text). The $1/\Delta f$ noise is centered around the carrier frequency $\Omega_{\mathrm{rf}} = 2\pi \times 30\,\mathrm{MHz}$. The motional sidebands at $\Omega_{\mathrm{rf}} \pm \omega_{\mathrm{x}}$ can be excited by noise at these frequencies (29 and 31 MHz).

η_{l} is capacitively coupled to nearby DC electrodes. If the DC electrode has similar electrical characteristics as the thin film RF electrode considered above, then this capacitive coupling would produce a $1/\Delta f$ voltage noise on the DC electrode proportional to η_{l}^2. For instance, consider a ground electrode which is directly beneath a trapped ion. This could be the case in a planar electrode ring trap or planar electrode linear trap. If it carries a substantial amount of the RF current ($\eta_{\mathrm{l}} = 1/3$) and has a characteristic distance $D = 100\,\mu\mathrm{m}$, this would result in an electric-field noise of approximately $10^{-8}\,\mathrm{V}^2/\mathrm{m}^2\mathrm{Hz}$ at the motional sidebands. This would give rise to a heating rate of about 1000 phonons/sec for $^{40}\mathrm{Ca}^+$.

Another concern is that stray DC fields and the excess micromotion they produce can result in ion heating from noise on the RF electrodes as it is pushed out of the RF null as explained above (see sec. E.5.1). For instance, if the ion is displaced 1μm from the RF null and the quadrupole characteristic distance D_{Q} of the trap is $100\,\mu\mathrm{m}$ (see eq. E.19), then the electric-field noise at the ion would be approximately $10^{-11}\,\mathrm{V}^2/\mathrm{m}^2\mathrm{Hz}$ at the motional side bands. This would produce a heating rate of only about 1 phonon/sec for $^{40}\mathrm{Ca}^+$, but would increase as the trap was scaled down as $1/d^4$.

E.6.3 $1/f$ **Noise with RF Excitation**

RF excitation has also been observed to produce a low frequency $1/f$ type noise, even though there is no current at DC [316–318]. This is not expected from just a fluctuation in resistance, as described above, and would require an alternative model [319]. The value of this RF excited $1/f$ noise is typically 1 or 2 orders of magnitude smaller than the level of $1/\Delta f$ noise. However, it could still be a source of concern, because of the much larger response of the ion at the motional frequency ω_x compared to the motional sidebands $\Omega_{rf} \pm \omega_x$. Continuing with the thin film example in this section, as shown in figure E.4, any low frequency noise present in the system at $\omega_x = 2\pi \times 1\,\mathrm{MHz}$ could be the dominant source of heating, because of the much larger heating response (65 dB) of the ion due to noise at its motional frequency (compared to the motional sidebands).

Flicker noise on thin clean metal films could be a concern for ion trap experiments. Furthermore, it may be that operation of the trap embeds ions into the metal film [320], so that the structure of the solid changes in a way that the Hooge parameter is much larger than that assumed above. So that if the trap exhibits large amounts of ion heating after being operated for a long time, testing for flicker noise should be done. The usual technique for measuring flicker noise requires that there are two connections on opposite sides of the conductor that is being tested for flicker noise. By measuring the voltage noise in the presence of a DC excitation current, the flicker noise can be measured. Some electrodes in surface traps, such as large ground planes and RF electrodes could be connected so that their are two available connections so as to test for flicker noise.

E.7   **Potential Drift**

Often seen in ion trap systems is a long term drift of the static potential near the metal electrodes such as described in figure 6.15 or documented by other researchers [214–218]. This is most likely caused by a change in the work function on parts of the surface of the electrodes due to an uneven coating of the electrode with a new material, such as from the trap loading oven.

For the trap array Folsom, a voltage drift was observed and was associated with increased heating, which was measured by a drop of the uncooled lifetime from seconds to hundreds of ms (see sec. 6.5). This increase in heating along with a voltage drift has also been observed in other experiments [214, 215]. The cause of this increased heating is not yet fully understood, though some have proposed that fluctuating patch potentials may be the cause as described below in section E.8. However, if the excess micromotion associated with the displacement of the ion from the RF null due to static fields is not compensated for, then excess heating will occur (see sec. E.5.1). It may be that some other mechanism also contributes to the increased heating associated with the static potential drift, such as flicker noise (see sec. E.6) or space charge (described below in section E.9).

E.8 Patch Potentials

The noise sources discussed in the previous section have been analyzed under the assumption that the whole trap electrode can be described as an ideal equipotential, generally resulting in a $\sim D^{-2}$ scaling of the noise with respect to the characteristic distance. This assumption is not in general true for real metallic surfaces, where regions of different crystal orientation or adsorbed atoms and compounds lead to local variations of the potential [321]. These so-called patch potentials play an important role in many areas of physics and represent, for example, an experimental limitation for precision measurements of the Casimir-Polder force between closely spaced metallic plates [322, 323] or gravity tests with charged elementary particles [324, 325]. It was first suggested by Turchette et al. [132] that fluctuating patch potentials on the electrodes could also be the source of the unexpectedly large heating rates observed in some ion trap experiments. In Turchette's original work, it was showed that for a simplified spherical trap geometry, and in the limit of the patches being small, the existence of local rather than extended voltage fluctuations leads to a d^{-4} scaling and therefore a strong enhancement of heating rates for small trap dimensions. Subsequent studies have investigated in more detail how the distance scaling is affected by finite patch sizes [326, 327] and by the electrode geometry [327] on the distance scaling. While those models make no predictions regarding the frequency or temperature dependence of the noise, they provide valuable predictions for distinguishing between, for example, technical noise sources leading to global fluctuations of the electrode voltage and noise sources related to microscopic processes on the electrode surface.

E.8.1 Origin of Patch Potentials

The term "patch-potential" refers quite generally to a local variation of the potential on an otherwise homogeneous, biased electrode surface. Different mechanisms are known to produce such microscopic potential variations. Most commonly patch potentials are attributed to regions of different crystal orientation and surface adsorbates [321]. For a clean and regular surface the otherwise homogeneous density of the electrons inside the metal is distorted at the surface, which creates an effective dipole layer at the metal-air interface. This dipole layer changes the work function, W, of the electrode by $\Delta W = e\Delta\Phi$. Here e is the charge of the electron and $\Delta\Phi$ is the patch potential, which is related to the dipole moment per unit area, $\mathcal{P}$, by [44],

$$\Delta\Phi = \mathcal{P}/\epsilon_0, \qquad (E.27)$$

The value of $\mathcal{P}$ depends on the surface properties, in particular on the relative orientation of the crystal lattice and the surface. Therefore, small regions of different crystal orientation can lead to variations of $\Delta\Phi$ over microscopic distances. A similar effect arises from adsorbed atoms and molecules, which are polarized when approaching the surface and form additional dipole layers. The static potentials of metallic surfaces have been measured using various methods. From thermionic-emission-current experiments, it is known that the work function of metal surfaces can vary by several tenths of a volt, depending on the crystal orientation [321]. On gold surfaces patch potentials with sizes ranging from $10\,\text{nm}$ to $10\,\mu\text{m}$, and $\Delta\Phi \sim \text{meV}$ have been measured using Kelvin probes [328, 329]. The surface dipoles created by adsorbates can be directly observed on the level of single atoms in precision experiments with cold trapped atoms. For example, Obrecht et al. [330] utilized a magnetically trapped BEC to measure the electric-field distribution emanating from a cluster of Rb atoms adsorbed on various surfaces. The induced dipole moment measured at $\mu \sim 5\,\mathrm{D}$ (debye) per atom is consistent with theoretical predictions for alkaline atoms absorbed on metallic surfaces. ($1\,\mathrm{D} \approx 3.33 \times 10^{-30}\,\mathrm{Cm}$.) Finally, trapped ions have been used to investigate laser-induced surface dipoles [175] and the long-term variations of stray electric fields over several months [214–218].

While static patch fields on metal surfaces are relatively well understood, little is known about their fluctuations, in particular in the MHz frequency regime of interest. It should also be noted that if the surface of an electrode is not geometrically smooth, fluctuating patch fields could also arise from a local enhancement of Johnson noise. For instance, if there is a small cylindrical cone with a spherical end-tip protruding above the surface, its resistance is given by Deslauriers et al. [242],

$$R = \frac{\rho_e}{\pi r_0 \tan\theta} \tag{E.28}$$

where ρ_e is the bulk resistivity, r_0 is the end-tip radius, and θ is the angle of the cone. Therefore, a small crystal of gold with a radius of a few nanometers at the tip, growing from the surface of an ion trap, could be equivalent to a nanometer-sized electrode having a resistance of several ohms. If the surface is populated with patches of a poor conductor, such as an oxide [175, 195], then small patches with orders-of-magnitude higher resistances could be expected.

E.9 Space Charge

The model of the ion-trap experiment assumes that the vacuum is neutral, but if some mechanism allows for charge to fill the vacuum then these charges could cause electric-field noise at the location of the ion trap. A common source of charge in vacuum which has been documented in ion traps [202] is electron emission from electrode surfaces. The exact mechanisms involved in electron emission could be field or thermal emission [331], or photoelectric emission [332], as ion traps can have sharp points, rough electrode surfaces, high voltages, locally hot electrodes and short-wavelength laser light. What the exact mechanism is would depend on each experiment, but the effect would be similar. Electrons escaping the surface of a cathodic electrode would follow the field lines created by the high-voltage trap drive and terminate at the anode. Secondary effects from the electrode bombardment would likely cause further disturbance, which we do not consider here.

Under typical ion-trap conditions, it would take ~ 100 picoseconds or less for the electron to reach the anode, so that the RF trap drive is essentially quasi-static during the process. Depending on where the electron escaped, it will go closer to the ion and then further away, following the arc of the electric-field line. Without solving for the exact electrostatic field of a particular trap geometry, it is interesting to do an analysis of a single electron flying through an ion trap at roughly a distance d from the ion. Taking the Coulomb force from any escaped electrons to be roughly modelled by a Gaussian pulse of temporal width $\tau_e = 100\,\text{ps}$ and electric-field amplitude of $q/(4\pi\epsilon_0 d^2)$, a Fourier analysis shows that the single sided energy spectral density $S_{\text{EE}}^{(\text{SC})}$ of the electric field noise during the electron emission is flat out to approximately $\nu_e = 1\,\text{GHz}$, with a height of

$$S_{\text{EE}}^{(\text{SC})} = 2\pi \left(\frac{q\tau_e}{4\pi\epsilon_0 d^2} \right)^2 \tag{E.29}$$

where q is the elementary charge. Considering the electron emission in a time window, $t_W = 1\,\text{s}$, the power spectral density (PSD) $S_{\text{E}}^{(\text{SC})}$ due to an electron emission rate of N electrons per second is then

$$S_{\text{E}}^{(\text{SC})} = \frac{2\pi N}{t_W} \left(\frac{q\tau_e}{4\pi\epsilon_0 d^2} \right)^2. \tag{E.30}$$

Using the above values, each electron would give an electric-field noise, $S_{\text{E}}^{(\text{SC})}$, roughly of order $10^{-21}\,\text{V}^2/\text{m}^2\text{Hz}$ for $d = 100\,\mu\text{m}$. This type of field noise is essentially due to the shot noise [333] of the electron-emission current. At an average emission current of $1\,\text{nA}$, a PSD of the electric-field noise of order $10^{-11}\,\text{V}^2/\text{m}^2\text{Hz}$ might be expected. Such an electron-emission current and associated field noise could be detected and engineered away. Though, it demonstrates the importance of checking for electron-emission in ion traps. The coating of the electrodes with low work-function materials, could considerably enhance electron-emission and increase the field noise in the vicinity of the ion.

It should be noted however that many electron emission experiments have seen that electron emission noise follows a low-frequency periodic or oscillatory nature, which is not white noise in character [334, 335], which would complicate the above analysis. It is also possible that the time-dependent nature of the trap drive at frequency Ω_{rf} could give rise to periodic field-induced electron-emission currents at frequency Ω_{rf}. The electron-emission current would then resemble a regular series of pulses, each of width τ_e, with an average temporal spacing at the period of the trap drive $T = 2\pi/\Omega_{\text{rf}}$. The above estimate would still be valid at frequencies well below the trap drive frequency Ω_{rf}, since the shot noise characteristics would dominate. However, the correlation of the electron-emission current at the trap drive frequency would enhance the field noise at harmonics of the trap drive frequency. In all cases, these point charges emanating from the surface of the electrodes would give give rise to distance scaling $\beta = 4$, as seen from equation (E.30), as long as all other operating parameters remain constant.

E.10 Spontaneous and Stimulated Emission of Trap Drive Photons

The ion in a quadrupole ion trap (QIT) also acts as a dipole antenna. It couples the trap-drive electric field, because of its motion, to the vacuum field as well as any ambient fields. Coupling to the vacuum field is termed spontaneous emission and coupling to ambient fields is termed stimulated emission. If the ion emits a photon, it can gain or lose energy depending on the sign of the *time averaged* $\overline{W}$ [75], which is given by

$$\overline{W} = \frac{1}{T} \int_t^{t+T} W(t)\, dt, \tag{E.31}$$

where T is the period of the trapped ion's periodic trapping potential $m_I W(t) x^2 / 2$ given in the Hamiltonian in equation 2.56. This is not a violation of energy, since the trapping fields supply the energy for the emitted RF photons, and the heating of the ion is the recoil of said photons. The exponential heating rate to lowest order is given by Glauber [75] as

$$\bar{n}(t) = n_0 \exp\left[-\frac{q^2}{6\pi\epsilon_0 m_I c^3} \overline{W} \right], \tag{E.32}$$

where $\bar{n}(t)$ is the expected value of the phonon number, n_0 is the initial phonon number and c is the speed of light. For traps with only a periodic component of the trapping potential, such as QITs without DC fields, the time-averaged potential $\overline{W}$ can be 0, so that the next higher order term is important which gives a linear heating rate to lowest order in q_x of

$$\bar{n}(t) = n_0 + \frac{q^2}{6\pi\epsilon_0 c^3} \left[\omega_x^2 + q_x \frac{(2\Omega_{rf} + \omega_x)^3}{2\omega_x} \right] t. \tag{E.33}$$

For spontaneously emitted RF photons, and for $^{40}\text{Ca}^+$ trapped in a linear trap with an axial frequency of $1\,\text{MHz}$, equation E.32 gives an exponential heating rate with a time constant of approximately 10^{15} seconds. This is because the DC fields would give a non-zero $\overline{W}$ proportional to the axial trapping frequency squared. For a point quadrupole ion trap without DC fields, the time-averaged $\overline{W}$ would be zero. And equation E.33 can be used with a trapping frequency $\omega_x = (2\pi \times 1\,\text{MHz}$ and a drive frequency $\Omega_{rf} = 30 \times 2\pi\,\text{MHz}$, to give a linear heating rate of roughly 10^{-13} phonons/s.

However the trap is not operating in a perfect electromagnetic vacuum, so stimulated emission must be taken in to account. Stimulated emission is proportional to the photon occupation number n, which for black-body radiation has an average value $\bar{n} = k_B T / (\hbar\omega)$, where ω corresponds to the stimulated emission photon frequency. The primary photon frequency of concern for the linear trap is ω_x, and for the point trap it is $\Omega_{rf} + \omega_x$. For a trap drive frequency of $\Omega_{rf} = 2\pi \times 30\,\text{MHz}$ and a motional frequency $\omega_x = 2\pi \times 1\,\text{MHz}$, the time constant for the linear trap in the presence of black-body radiation is then roughly $10^{-8}\,\text{s}$ and the heating rate for the point trap without DC fields is roughly 10^{-8} phonons/s.

As mentioned above in the section E.1 on coupling to fields in free space, the environment of a typical laboratory has much more electromagnetic noise than expected from the ideal of black-body radiation. Since electromagnetic noise can also stimulate emission, the external noise factor F_a, which measures the noise power relative to black-body radiation must also be taken into account. At a frequency near $1\,\text{MHz}$, the outdoor noise factor near manmade structures is approximately $F_a = 80\,\text{dB}$, so that the expected stimulated emission rate is 10^8 more likely. The noise factor at $1\,\text{MHz}$ in an area filled with electronics equipment could easily exceed this value by $40\,\text{dB}$ [336]. If ambient noise is such that $F_a = 80\,\text{dB}$, the exponential time constant would be reduced to approximately 1 second for the linear trap. For the linear heating rate of a point quadrupole ion trap without DC fields, one might expect 1 phonons/s. Excess micromotion due to stray DC fields could enhance this effect, and would remove the assumption that point ion traps could be operated with $\overline{W} = 0$. The exponential heating could be of concern and points to the need to control the electric-field noise at the trap for very long uncooled lifetimes of many minutes or hours. It should be a simple matter of measuring the spectral noise density at the trap drive frequencies to make sure that the noise factor is not so large as to limit the experiment. Appropriate RF shielding could then be installed if needed so as to reduce the noise factor.

Appendix F

Fabrication Steps for Ziegelstadl

In this appendix, the fabrication process chains for the Ziegelstadl microtraps (see sec. 7.1) are given. The fabrication steps were performed at the Fachhochschule Vorarlberg in Dornbirn, Austria in Prof. Johannes Edlinger's Microtechnology research group (sputtering, resist coating and development, and etching) and at the Universität Innsbruck in Prof. Rainer Blatt's ion trapping research group (evaporation and gold bonding).

The thin-film metal lithography steps to make Ziegelstadl are generally described as follows: First, a planar Pyrex™ substrate of either 100 mm or 150 mm diameter was metalized using a sputtering process. Then a photoresist was spin coated onto the substrate. The photoresist was then developed using a chrome mask to reveal the metal which was to be etched away. Then the etching process was performed. There are two ways the etching was done. One way was to use a bath of acid (wet etch), and the other was to use a plasma (dry etch). After the first etching process, the first resist could be removed to reveal a metal structure. Then a silica insulating layer SiO_2 was then applied using a sputtering process. After applying a photoresist and developing it, the glass was then etched using a dry etch process. Then the resist could be removed to reveal a patterned dielectric layer. Finally the top electrode metalization could be done. One way to metalize and pattern the electrodes has already been described (by metalization, resist and etching). Another way is to first apply the photoresist and develop it, and then to evaporate metal into the exposed areas. After this, the resist is cleaned away, leaving behind a thin film metal structure. This is called a lift-off process. After all the layers were patterned, a saw was used to cut up the wafers (wafer dicing) into small chips, which could be then glued to a circuit board and gold bonded to traces on the circuit board. In this way, a two layer circuit, with vias between the two layers could be fabricated and connectorized so as to trap ions.

There were four different processes used to make Ziegelstadl. The first is detailed in figure F.1. The process is detailed with the first step on the bottom of the figure and the last step at the top of the figure. Starting with the first step at the bottom of the figure, the sputtering of the metal is described. It has electrodes composed of a sandwich of 50 nm titanium, 1 µm copper and 50 nm titanium (Ti/Cu/Ti). The next step shows the resist type (AZ1518) and its thickness (1.8µm). The next step describes the wet etch process used on the bottom metal layer. The Ti was etched with a 1% solution of HF acid and the Cu was etched with a commercial copper etch

(Atotech). Then the cleaning process is depicted to remove the resist (with Acetone). Next the SiO_2 deposition is done by sputtering with a thickness of 1µm. Next the resist and etching steps for the insulation layer are described. The cleaning steps before the final metalization are then given. The final metalization and patterning is then shown. The purpose of the titanium is to protect the copper from oxidation as well as promote adhesion to the silica insulating layer or Pyrex™ substrate. The titanium however proved to make the bond pads, which were already quite small, impossible to bond with gold wire. Various approaches were tried to make the bond pads bondable. The first was to etch away the titanium protection layer, but the copper electrodes then quickly oxidized, so that only a fraction of the pads could then be bonded successfully. The resulting copper surface was also a source of concern for trapping ions, especially after the high heating rates were measured in Folsom. For these reasons, a fabrication run, using only titanium and gold electrodes was attempted.

Figure F.2 shows the fabrication steps for creating a trap with surface electrodes made of titanium (700 nm) with a covering of 300 nm of gold (Ti/Au). The steps are similar to that given above. The lower layer was still composed of the Ti/Cu/Ti sandwich described above. However, the patterning of the top metal layer was performed differently. The lower layer was wet etched, the glass insulating layer was dry etched, and the top electrode layer used a lift-off process to pattern the metal. Unfortunately, the tension created by the titanium evaporation as it cooled, made the top metal layer very weak and prone to peeling. Only a minor fraction (about 5%) of the traps on the wafer had all the electrodes intact after the lift-off process. Of these traps, gold bonding was attempted, but the tension of the thick layer of titanium made the pads peel off the substrate when bonding was attempted. For these reasons, a new version of Ziegelstadl was designed with larger bond pads and a new metalization process.

Figure F.3 and F.4 show slightly different steps (differing resist thicknesses) for producing Ziegelstadl v1.1. This trap used electrodes made by sputtering titanium with a thickness of 1 µm for both the top and bottom layers. All etching was performed with a plasma process. In order to make the trap easily gold-bondable and to provide a noble metal surface for ion trapping, a final metalization was done via a lift-off process of titanium 2 nm and 300 nm of gold. The trap proved to be easily bondable. However, this time the glass insulating layer did not function, as all electrical tests failed at 1 V or less. It is believe this is because the plasma etching of the top layer created a large voltage which broke-down and destroyed the functionality of the glass dielectric [337, 338]. One method to avoid this, would be to short the two electrode layers during fabrication and then cut away these connections during wafer dicing.

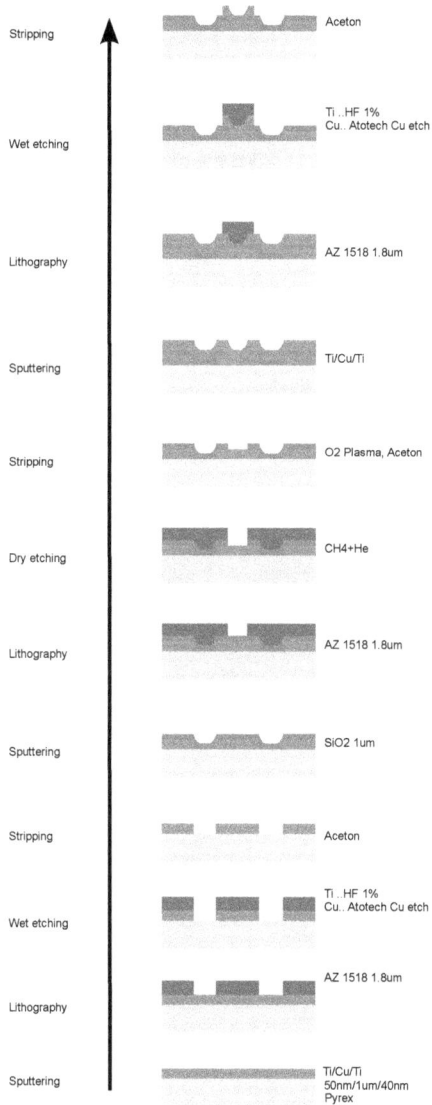

Figure F.1: Fabrication steps (ordered from bottom to top) for Ziegelstadl v1.0 Cu electrodes.

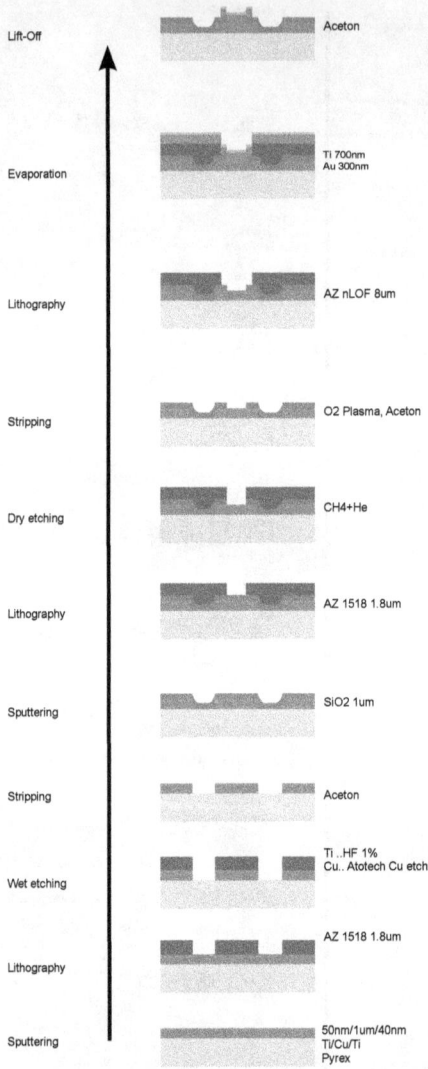

Figure F.2: Fabrication steps (ordered from bottom to top) for Ziegelstadl v1.0 Au/Ti electrodes.

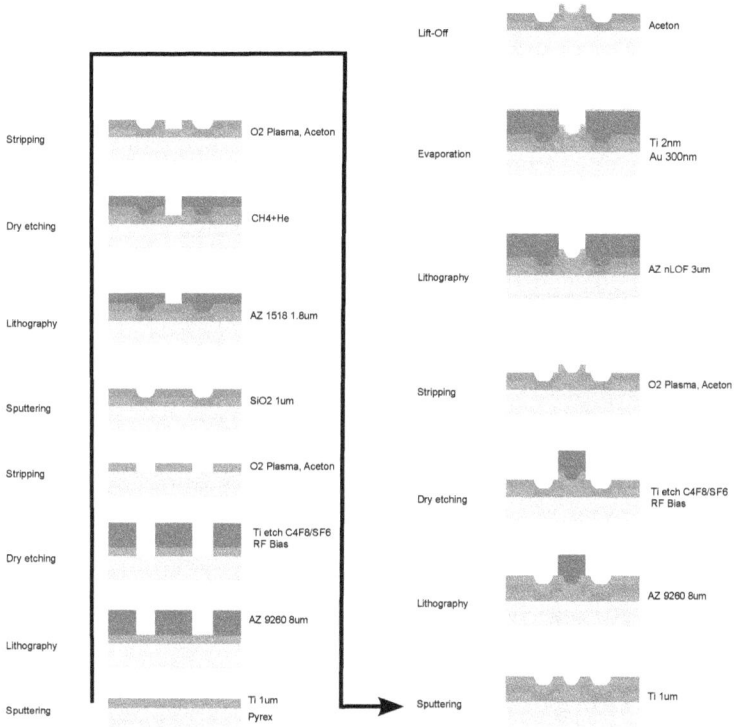

Figure F.3: Fabrication steps (ordered from bottom to top and left to right) for Ziegelstadl v1.1 Wafer #1.

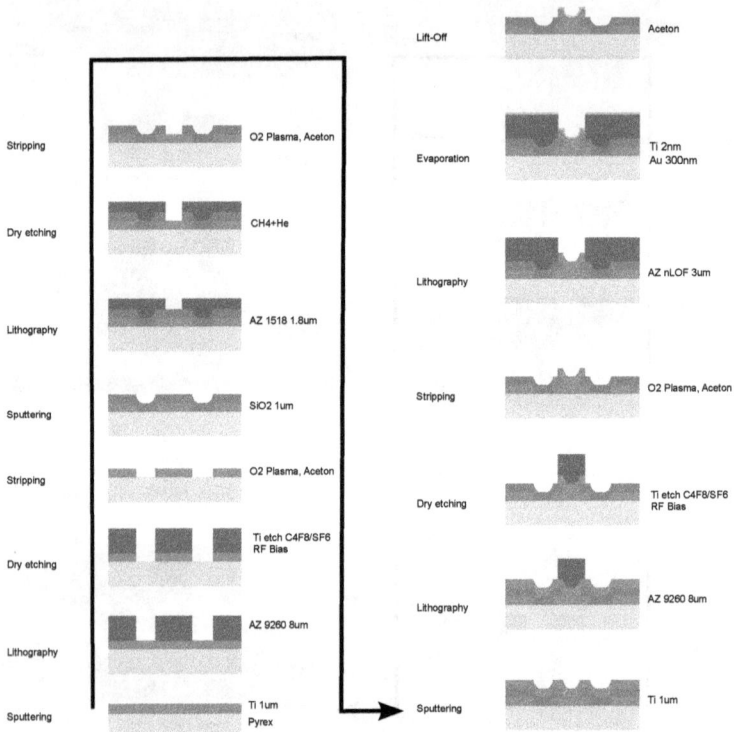

Figure F.4: Fabrication steps (ordered from bottom to top and left to right) for Ziegelstadl v1.1 Wafer #2.

Bibliography

[1] R. P. FEYNMAN. Simulating physics with computers. *International Journal of Theoretical Physics*, **21**: 467–488, 1982. DOI: 10.1007/BF02650179 (see pp. iii, v, 7)

[2] D. DEUTSCH. Quantum Theory, the Church-Turing Principle and the Universal Quantum Computer. *Proceedings of the Royal Society A: Mathematical, Physical and Engineering Sciences*, **400**: 97–117, 1985. DOI: 10.1098/rspa.1985.0070 (see pp. iii, v, 6, 8, 46)

[3] P. J. MOHR, B. N. TAYLOR, and D. B. NEWELL. CODATA recommended values of the fundamental physical constants: 2010. *Reviews of Modern Physics*, **84**: 1527–1605, 2012. DOI: 10.1103/RevModPhys.84.1527 (see p. xxi)

[4] M. J. BEESON. „The Mechanization of Mathematics" in: *Alan Turing: Life and Legacy of a Great Thinker* ed. by C. TEUSCHER. New York: Springer, 2004. 82 (see p. 1)

[5] A. CHURCH. An Unsolvable Problem of Elementary Number Theory. *American Journal of Mathematics*, **58**: 345–363, 1936. (see p. 1)

[6] A. M. TURING. On Computable Numbers, with an Application to the Entscheidungsproblem. *Proceedings of the London Mathematical Society*, **s2-42**: 230–265, 1937. DOI: 10.1112/plms/s2-42.1.230 (see p. 1)

[7] B. ROSSER. An Informal Exposition of Proofs of Gödel's Theorems and Church's Theorem. *The Journal of Symbolic Logic*, **4**: 53–60, 1939. (see p. 1)

[8] F. W. KISTERMANN. Blaise Pascal's Adding Machine: New Findings and Conclusions by Friedrich W. Kistermann. *IEEE Annals of the History of Computing*, **20**: 69, 1998. DOI: 10.1109/85.646211 (see p. 2)

[9] N. J. LEHMANN. Neue Erfahrungen zur Funktionsfähigkeit von Leibniz' Rechenmaschine. *Studia Leibnitiana*, **25**: 174–188, 1993. (see p. 2)

[10] M. A. AXTELL. Garbage Can Music!: Rube Goldberg's Three Careers. *Columbia Journal of American Studies*, **7**: 30, 2006. (see p. 2)

[11] T. YOUNG. The Bakerian Lecture: On the Theory of Light and Colours. *Philosophical Transactions of the Royal Society of London*, **92**: 12–48, 1802. (see p. 2)

[12] G. I. TAYLOR. Interference fringes with feeble light. *Proceedings of the Cambridge Philosophical Society*, **15**: 114, 1909. (see p. 2)

[13] P. A. M. DIRAC. A new notation for quantum mechanics. *Mathematical Proceedings of the Cambridge Philosophical Society*, **35**: 416–418, 1939. DOI: 10.1017/S0305004100021162 (see p. 3)

[14] R. C. HANNA. Polarization of annihilation radiation. *Nature*, **162**: 332, 1948. DOI: 10. 1038/162332a0 (see p. 4)

[15] C. S. WU and I. SHAKNOV. The angular correlation of scattered annihilation radiation. *Physical Review*, **77**: 136–137, 1950. DOI: 10.1103/PhysRev.77.136 (see p. 4)

[16] J. BELL. „Speakable and unspeakable in quantum mechanics" in: *Collected Papers on Quantum Philosophy*. 1987. (see p. 4)

[17] A. EINSTEIN, B. PODOLSKY, and N. ROSEN. Can quantum-mechanical description of physical reality be considered complete? *Physical Review*, **47**: 777, 1935. DOI: 10.1103/ PhysRev.47.777 (see p. 4)

[18] J. BELL. On the Einstein Podolsky Rosen Paradox. *Physics*, **1**: 195, 1964. (see p. 4)

[19] J. F. CLAUSER, M. A. HORNE, A. SHIMONY, and R. A. HOLT. Proposed Experiment to Test Local Hidden-Variable Theories. *Physical Review Letters*, **23**: 880–884, 1969. DOI: 10.1103/PhysRevLett.23.880 (see p. 5)

[20] M. A. HORNE, A. SHIMONY, and A. ZEILINGER. Two-particle interferometry. *Physical Review Letters*, **62**: 2209–2212, 1989. DOI: 10.1103/PhysRevLett.62.2209 (see p. 5)

[21] R. GHOSH and L. MANDEL. Observation of nonclassical effects in the interference of two photons. *Physical Review Letters*, **59**: 1903–1905, 1987. DOI: 10.1103/PhysRevLett.59. 1903 (see p. 5)

[22] C. SHANNON. Programming a computer for playing chess. *Philosophical Magazine*, **41**: 256, 1950. DOI: 10.1080/14786445008521796 (see p. 6)

[23] J. N. NEWMAN. *Marine Hydrodynamics*. Cambridge, Massachusetts: The MIT Press, 1977. (see p. 7)

[24] M. LEE, N. MALAYA, and R. D. MOSER. „Petascale direct numerical simulation of tur- bulent channel flow on up to 786K cores" in: *Proceedings of the International Conference for High Performance Computing, Networking, Storage and Analysis on - SC '13*. New York, New York, USA: ACM Press, 2013. 1–11 DOI: 10.1145/2503210.2503298 (see p. 7)

[25] C. E. TANSLEY and D. P. MARSHALL. Flow past a cylinder on a β plane, with application to Gulf Stream separation and the Antarctic Circumpolar Current. *Journal of Physical Oceanography*, **31**: 3274–3283, 2001. DOI: 10.1175/1520-0485(2001)031<3274:FPACOA> 2.0.CO;2 (see p. 7)

[26] H. SUTTER. The Free Lunch Is Over. *Dr. Dobbs Journal*, **30**: 2009. (see p. 7)

[27] S. LLOYD. Almost Any Quantum Logic Gate is Universal. *Physical Review Letters*, **75**: 346–349, 1995. DOI: 10.1103/PhysRevLett.75.346 (see pp. 7, 44)

[28] I. BULUTA and F. NORI. Quantum Simulators. *Science*, **326**: 108–111, 2009. DOI: 10. 1126/science.1177838 (see pp. 7, 88)

[29] E. ISING. Beitrag zur Theorie des Ferromagnetismus. *Zeitschrift für Physik*, **31**: 253–258, 1925. DOI: 10.1007/BF02980577 (see pp. 7, 8)

[30] R. MOESSNER, S. L. SONDHI, and P. CHANDRA. Two-Dimensional Periodic Frustrated Ising Models in a Transverse Field. *Physical Review Letters*, **84**: 4457–4460, 2000. DOI: 10.1103/PhysRevLett.84.4457 (see pp. 7, 8)

[31] J. W. BRITTON, B. C. SAWYER, A. C. KEITH, C.-C. J. WANG, J. K. FREERICKS, H. UYS, M. J. BIERCUK, and J. J. BOLLINGER. Engineered two-dimensional Ising interactions in a trapped-ion quantum simulator with hundreds of spins. *Nature*, **484**: 489–492, 2012. DOI: 10.1038/nature10981 (see p. 8)

[32] A. W. HARROW, A. HASSIDIM, and S. LLOYD. Quantum Algorithm for Linear Systems of Equations. *Physical Review Letters*, **103**: 150502, 2009. DOI: 10.1103/PhysRevLett.103.150502 (see p. 8)

[33] X.-D. CAI, C. WEEDBROOK, Z.-E. SU, M.-C. CHEN, M. GU, M.-J. ZHU, L. LI, N.-L. LIU, C.-Y. LU, and J.-W. PAN. Experimental Quantum Computing to Solve Systems of Linear Equations. *Physical Review Letters*, **110**: 230501, 2013. DOI: 10.1103/PhysRevLett.110.230501 (see p. 8)

[34] P. W. SHOR. „Algorithms for Quantum Computation: Discrete Logarithms and Factoring" in: *35th Annual Symposium on Foundations of Computer Science*. IEEE Computer Society Press, 1994. 124 DOI: 10.1109/SFCS.1994.365700 (see p. 9)

[35] M. KUMPH, M. BROWNNUTT, and R. BLATT. Two-dimensional arrays of radio-frequency ion traps with addressable interactions. *New Journal of Physics*, **13**: 073043, 2011. DOI: 10.1088/1367-2630/13/7/073043 (see pp. 9, 76)

[36] E. MARTÌN-LÓPEZ, A. LAING, T. LAWSON, R. ALVAREZ, X.-Q. ZHOU, and J. L. O'BRIEN. Experimental realization of Shor's quantum factoring algorithm using qubit recycling. *Nature Photonics*, **6**: 773, 2012. DOI: 10.1038/nphoton.2012.259 (see p. 11)

[37] A. BLAIS, R.-S. HUANG, A. WALLRAFF, S. GIRVIN, and R. SCHOELKOPF. Cavity quantum electrodynamics for superconducting electrical circuits: An architecture for quantum computation. *Physical Review A*, **69**: 062320, 2004. DOI: 10.1103/PhysRevA.69.062320 (see p. 11)

[38] D. D. AWSCHALOM, L. C. BASSETT, A. S. DZURAK, E. L. HU, and J. R. PETTA. Quantum spintronics: engineering and manipulating atom-like spins in semiconductors. *Science (New York, N.Y.)*, **339**: 1174–9, 2013. DOI: 10.1126/science.1231364 (see p. 11)

[39] I. BLOCH. Quantum coherence and entanglement with ultracold atoms in optical lattices. *Nature*, **453**: 1016–22, 2008. DOI: 10.1038/nature07126 (see p. 11)

[40] A. M. STEANE. The ion trap quantum information processor. *Applied Physics B: Lasers and Optics*, **642**: 623–642, 1997. DOI: 10.1007/s003400050225 (see pp. 11, 28, 56, 57, 60, 67)

[41] F. M. PENNING. Die Glimmentladung bei niedrigem Druck zwischen koaxialen Zylindern in einem axialen Magnetfeld. *Physica*, **3**: 873–894, 1936. DOI: 10.1016/S0031-8914(36)80313-9 (see p. 12)

[42] W. PAUL, O. OSBERGHAUS, and E. FISCHER. Ein Ionenkäfig. *Forschungsberichte des Wirtschafts- und Verkehrsministeriums Nordrhein Westfalen*, **415**: 1958. (see pp. 12, 14)

[43] K. H. KINGDON. A Method for the Neutralization of Electron Space Charge by Positive Ionization at Very Low Gas Pressures. *Physical Review*, **21**: 408–418, 1923. DOI: 10.1103/PhysRev.21.408 (see p. 12)

[44] J. D. JACKSON. *Classical Electrodynamics*. 3rd New York: Wiley, 1999. (see pp. 13, 63, 193)

[45] A. PAULI *Classical control of an ion in a surface trap* 2011 (see p. 13)

[46] S. EARNSHAW. On the nature of the molecular forces which regulate the constitution of the luminiferous ether. *Transactions of the Cambridge Philosophical Society*, **7**: 97, 1842. (see p. 14)

[47] R. F. WUERKER, H. SHELTON, and R. V. LANGMUIR. Electrodynamic Containment of Charged Particles. *Journal of Applied Physics*, **30**: 342–349, 1959. DOI: http://dx.doi.org/10.1063/1.1735165 (see p. 14)

[48] E. FISCHER. Die dreidimensionale Stabilisierung von Ladungsträgern in einem Vierpolfeld. *Zeitschrift für Physik*, **156**: 1–26, 1959. DOI: 10.1007/BF01332512 (see p. 14)

[49] W. PAUL. Electromagnetic traps for charged and neutral particles. *Reviews of Modern Physics*, **62**: 531, 1990. (see pp. 14, 131)

[50] E. MATHIEU. Mémoire sur Le Mouvement Vibratoire d'une Membrane de forme Elliptique. *Journal de Mathématiques Pures et Appliquées*, **13**: 137, 1868. (see p. 15)

[51] P. K. GHOSH. *Ion Traps*. Oxford University Press, 1995. (see pp. 15, 16, 35)

[52] R. G. DEVOE. Elliptical ion traps and trap arrays for quantum computation. *Physical Review A*, **58**: 910–914, 1998. DOI: 10.1103/PhysRevA.58.910 (see p. 15)

[53] G. W. HILL. On the part of the motion of the lunar perigee which is a function of the mean motions of the sun and moon. *Acta Mathematica*, **8**: 1–36, 1886. (see p. 15)

[54] W. MAGNUS and S. WINKLER. *Hill's Equation*. Dover Publications, Inc, 1979. (see p. 15)

[55] G. FLOQUET. Sur les équations différentielles linéaires à coefficients périodiques. *Annales de l'École Normale Supérieure*, **12**: 47, 1883. (see p. 15)

[56] W. MAGNUS *Infinite Determinants in the Theory of Mathieu's and Hill's Equations* tech. rep. 600 New York University, 1953 (see p. 16)

[57] N. W. MCLACHLAN. *Theory and application of Mathieu Functions*. Oxford University Press, 1947. (see p. 16)

[58] M. ABRAMOWITZ and I. A. STEGUN. *Handbook of mathematical functions (10th Printing)*. National Bureau of Standards, 1972. (see p. 16)

[59] D. J. BATE, K. DHOLAKIA, R. C. THOMPSON, and D. C. WILSON. Ion Oscillation Frequencies in a Combined Trap. *Journal of Modern Optics*, **39**: 305–316, 1992. DOI: 10.1080/09500349214550301 (see p. 16)

[60] H. DEHMELT. Radiofrequency Spectroscopy of Stored Ions I: Storage. *Advances in Atomic and Molecular Physics*, **3**: 53–72, 1968. DOI: 10.1016/S0065-2199(08)60170-0 (see p. 17)

[61] G. HERZBERG. *Atomic Spectra and Atomic Structure*. 2nd Prentice-Hall, Inc., 1944. (see p. 22)

[62] C. J. FOOT. „Atomic Physics" in: *Atomic Physics*. Oxford University Press, 2005. (see p. 22)

[63] F. SAUNDERS and H. RUSSELL. On the Spectrum of Ionized Calcium (Ca II). *The Astrophysical Journal*, **62**: 1, 1925. DOI: 10.1086/142908 (see pp. 22, 24)

[64] J. JIN and D. A. CHURCH. Precision lifetimes for the Ca^+ $4p^2P$ levels: Experiment challenges theory at the 1% level. *Physical Review Letters*, **70**: 3213 3216, 1993. DOI: 10.1103/PhysRevLett.70.3213 (see p. 24)

[65] P. A. BARTON, C. J. S. DONALD, D. M. LUCAS, D. A. STEVENS, A. M. STEANE, and D. N. STACEY. Measurement of the lifetime of the $3d^2D_{5/2}$ state in $^{40}Ca^+$. *Physical Review A*, **62**: 032503, 2000. DOI: 10.1103/PhysRevA.62.032503 (see p. 24)

[66] R. GERRITSMA, G. KIRCHMAIR, F. ZÄHRINGER, J. BENHELM, R. BLATT, and C. F. ROOS. Precision measurement of the branching fractions of the 4p $^2P_{3/2}$ decay of Ca II. *The European Physical Journal D*, **50**: 13–19, 2008. DOI: 10.1140/epjd/e2008-00196-9 (see p. 24)

[67] M. RAMM, T. PRUTTIVARASIN, M. KOKISH, I. TALUKDAR, and H. HÄFFNER. Precision Measurement Method for Branching Fractions of Excited P1/2 States Applied to Ca+40. *Physical Review Letters*, **111**: 023004, 2013. DOI: 10.1103/PhysRevLett.111.023004 (see p. 24)

[68] D. F. V. JAMES. Quantum dynamics of cold trapped ions with application to quantum computation. *Applied Physics B: Lasers and Optics*, **66**: 181–190, 1998. DOI: 10.1007/s003400050373 (see pp. 28, 181)

[69] J. P. HOME and A. M. STEANE. Electrode Configurations for Fast Separation of Trapped Ions. *Quantum Information and Computation*, **6**: 289, 2006. (see pp. 28, 69)

[70] J. HOME, D. HANNEKE, J. JOST, D. LEIBFRIED, and D. J. WINELAND. Normal modes of trapped ions in the presence of anharmonic trap potentials. *New Journal of Physics*, **13**: 073026, 2011. DOI: 10.1088/1367-2630/13/7/073026 (see pp. 28, 60)

[71] T. MONZ, P. SCHINDLER, J. T. BARREIRO, M. CHWALLA, D. NIGG, W. A. COISH, M. HARLANDER, W. HÄNSEL, M. HENNRICH, and R. BLATT. 14-Qubit Entanglement: Creation and Coherence. *Physical Review Letters*, **106**: 130506, 2011. DOI: 10.1103/PhysRevLett.106.130506 (see p. 29)

[72] L. S. BROWN. Quantum motion in a Paul trap. *Physical Review Letters*, **66**: 527–529, 1991. DOI: 10.1103/PhysRevLett.66.527 (see p. 29)

[73] D. LEIBFRIED, R. BLATT, C. MONROE, and D. J. WINELAND. Quantum dynamics of single trapped ions. *Reviews of Modern Physics*, **75**: 281, 2003. DOI: 10.1103/RevModPhys.75.281 (see pp. 29–31, 33, 39, 50, 132)

[74] M. COMBESCURE. A quantum particle in a quadrupole radio-frequency trap. *Annales de l'institut Henri Poincaré (A) Physique théorique*, **44**: 293–314, 1986. (see p. 29)

[75] R. J. GLAUBER. „The Quantum Mechanics of Trapped Wave Packets" in: *Proceedings of the International School of Physics "Enrico Fermi" Course 118 "Laser Manipulation of Atoms and Ions"* ed. by E. ARIMONDO, W. PHILLIPS, and F. STRUMIA. Società Italiana di Fisica, 1992. 643 (see pp. 29, 195, 196)

[76] L. I. SCHIFF. „Interaction Picture" in: *Quantum Mechanics*. 3rd McGraw-Hill Book Company, Inc, 1955. chap. Matrix For, 171 (see p. 32)

[77] D. J. WINELAND, C. MONROE, W. ITANO, B. E. KING, D. LEIBFRIED, D. M. MEEKHOF, C. J. MYATT, and C. S. WOOD. Experimental Primer on the Trapped Ion Quantum Computer. *Fortschritte der Physik*, **46**: 363–390, 1998. DOI: 10.1002/(SICI)1521-3978(199806)46:4/5<363::AID-PROP363>3.0.CO;2-4 (see pp. 33, 48, 50, 51, 55, 181)

[78] I. I. RABI, J. R. ZACHARIAS, S. MILLMAN, and P. KUSCH. A New Method of Measuring Nuclear Magnetic Moment. *Physical Review*, **53**: 318, 1938. DOI: 10.1103/PhysRev.53.318 (see p. 34)

[79] P. SCHINDLER, D. NIGG, T. MONZ, J. T. BARREIRO, E. MARTINEZ, S. X. WANG, S. QUINT, M. F. BRANDL, V. NEBENDAHL, C. F. ROOS, M. CHWALLA, M. HENNRICH, and R. BLATT. A quantum information processor with trapped ions. *New Journal of Physics*, **15**: 123012, 2013. DOI: 10.1088/1367-2630/15/12/123012 (see p. 35)

[80] S. T. GULDE, D. ROTTER, P. BARTON, F. SCHMIDT-KALER, R. BLATT, and W. HOGERVORST. Simple and efficient photo-ionization loading of ions for precision ion-trapping experiments. *Applied Physics B*, **73**: 861, 2001. DOI: 10.1007/s003400100749 (see pp. 35, 95)

[81] D. M. LUCAS, A. RAMOS, J. P. HOME, M. J. MCDONNELL, S. NAKAYAMA, J.-P. STACEY, S. C. WEBSTER, D. N. STACEY, and A. M. STEANE. Isotope-selective photoionization for calcium ion trapping. *Physical Review A*, **69**: 012711, 2004. DOI: 10.1103/PhysRevA.69.012711 (see pp. 35, 139)

[82] R. LECHNER *Photoionisation of 40 Ca with a frequency-doubled 422 nm laser and a 377 nm laser diode* 2010 (see p. 35)

[83] D. WINELAND and H. DEHMELT. „Proposed 10^{14} $\Delta V < V$ Laser Fluorescence Spectroscopy on Tl$^+$ Mono-Ion Oscillator III" in: *Bulletins of the American Physical Society.* vol. 20 1975. 637 (see p. 35)

[84] D. J. WINELAND, R. E. DRULLINGER, and F. L. WALLS. Radiation-Pressure Cooling of Bound Resonant Absorbers. *Physical Review Letters*, **40**: 1639–1642, 1978. DOI: 10.1103/PhysRevLett.40.1639 (see p. 35)

[85] G. WERTH, V. N. GHEORGHE, and F. G. MAJOR. *Charged Particle Traps II: Applications.* Berlin Heidelberg: Springer, 2009. (see p. 35)

[86] C. DOPPLER. Über das farbige Licht der Doppelsterne und einiger anderer Gestirne des Himmels. *Abhandlungen der königlich böhmischen Gesellschaft der Wissenschaften (V Folge, Bd. 2), in: Landespressebüro der Salzburger Landesregierung (Hrsg.): Christian Doppler-Leben und Werk. Schriftenreihe Serie Sonderpublikationen*, **5**: 1842. (see pp. 35, 37)

[87] F. DIEDRICH, J. C. BERGQUIST, W. ITANO, and D. J. WINELAND. Laser Cooling to the Zero-Point Energy of Motion. *Physical Review Letters*, **62**: 403, 1989. DOI: 10.1103/PhysRevLett.62.403 (see pp. 36, 170)

[88] S. STENHOLM. The semiclassical theory of laser cooling. *Reviews of Modern Physics*, **58**: 699–739, 1986. DOI: 10.1103/RevModPhys.58.699 (see p. 37)

[89] J. I. CIRAC, R. BLATT, P. ZOLLER, and W. D. PHILLIPS. Laser cooling of trapped ions in a standing wave. *Physical Review A*, **46**: 2668–2681, 1992. DOI: 10.1103/PhysRevA. 46.2668 (see p. 38)

[90] J. ESCHNER, G. MORIGI, F. SCHMIDT-KALER, and R. BLATT. Laser cooling of trapped ions. *Journal of the Optical Society of America B*, **20**: 1003, 2003. DOI: 10.1364/JOSAB. 20.001003 (see p. 38)

[91] R. LOUDON. *The Quantum Theory of Light*. Third Oxford University Press, 2000. (see p. 38)

[92] G. ALZETTA, A. GOZZINI, L. MOI, and G. ORRIOLS. An experimental method for the observation of r.f. transitions and laser beat resonances in oriented Na vapour. *Il Nuovo Cimento B Series 11*, **36**: 5–20, 1976. DOI: 10.1007/BF02749417 (see p. 39)

[93] E. ARIMONDO. „V Coherent Population Trapping in Laser Spectroscopy" in: *Progress in Optics* ed. by E. WOLF. vol. 35 Progress in Optics Elsevier, 1996. 257–354 DOI: http://dx.doi.org/10.1016/S0079-6638(08)70531-6 (see p. 39)

[94] I. SIEMERS. *Über die Dunkelresonanz im Anregungsspektrum eines einzelnen gespeicherten Ions*. PhD thesis. Universität Hamburg, 1991. (see pp. 39, 40)

[95] I. SIEMERS, M. SCHUBERT, R. BLATT, W. NEUHAUSER, and P. E. TOSCHEK. The "Trapped State" of a Trapped Ion-Line Shifts and Shape. *EPL (Europhysics Letters)*, **18**: 139, 1992. DOI: 10.1209/0295-5075/18/2/009 (see pp. 39, 40)

[96] D. J. BERKELAND. Destabilization of dark states and optical spectroscopy in Zeeman-degenerate atomic systems. *Physical Review A*, **65**: 033413, 2002. DOI: 10.1103/PhysRevA. 65.033413 (see pp. 40, 41)

[97] R. WYNANDS and A. NAGEL. Precision spectroscopy with coherent dark states. *Applied Physics B*, **68**: 1–25, 1999. DOI: 10.1007/s003400050581 (see p. 40)

[98] C. MONROE, D. M. MEEKHOF, B. E. KING, S. JEFFERTS, W. ITANO, D. J. WINELAND, and P. GOULD. Resolved-Sideband Raman Cooling of a Bound Atom to the 3D Zero-Point Energy. *Physical Review Letters*, **75**: 4011–4014, 1995. (see pp. 41, 170)

[99] P. Z. PEEBLES. *Communication System Principles*. Addison-Wesley Publishing Co., 1976. (see p. 41)

[100] C. ROOS, T. ZEIGER, H. ROHDE, H. C. NÄGERL, J. ESCHNER, D. LEIBFRIED, F. SCHMIDT-KALER, and R. BLATT. Quantum State Engineering on an Optical Transition and Decoherence in a Paul Trap. *Physical Review Letters*, **83**: 4713, 1999. DOI: 10.1103/PhysRevLett.83.4713 (see pp. 42, 170)

[101] M. A. NIELSEN and I. L. CHUANG. *Quantum Computation and Quantum Information*. Cambridge University Press, 2000. (see p. 43)

[102] M. RIEBE. *Preparation of Entangles States and Quantum Teleportation with Atomic Qubits*. PhD thesis. Universitäts Innsbruck, 2005. (see pp. 43, 48)

[103] P. SCHINDLER. *Quantum computation and simulation with trapped ions using dissipation*. PhD thesis. Universität Innsbruck, 2013. (see pp. 43, 47)

[104] J. I. CIRAC and P. ZOLLER. Quantum Computations with Cold Trapped Ions. *Physical Review Letters*, **74**: 4091–4094, 1995. DOI: 10.1103/PhysRevLett.74.4091 (see pp. 44, 67)

[105] A. S. SØRENSEN and K. MØLMER. Entanglement and quantum computation with ions in thermal motion. *Physical Review A*, **62**: 022311, 2000. DOI: 10.1103/PhysRevA.62.022311 (see pp. 44, 88)

[106] A. BERMUDEZ, P. O. SCHMIDT, M. B. PLENIO, and A. RETZKER. Robust trapped-ion quantum logic gates by continuous dynamical decoupling. *Physical Review A*, **85**: 040302, 2012. DOI: 10.1103/PhysRevA.85.040302 (see pp. 44, 47, 66)

[107] G. KIRCHMAIR, J. BENHELM, F. ZÄHRINGER, R. GERRITSMA, C. F. ROOS, and R. BLATT. Deterministic entanglement of ions in thermal states of motion. *New Journal of Physics*, **11**: 023002, 2009. DOI: 10.1088/1367-2630/11/2/023002 (see p. 44)

[108] J. I. CIRAC and P. ZOLLER. A scalable quantum computer with ions in an array of microtraps. *Nature*, **13**: 579, 2000. DOI: 10.1038/35007021 (see pp. 44, 61, 67, 87, 88)

[109] H. DEHMELT. „Proposed 10^{14} $\Delta V < V$ Laser Fluorescence Spectroscopy on Tl$^+$ Mono-Ion Oscillator II" in: *Bulletins of the American Physical Society*. vol. 20 1975. 60 (see p. 44)

[110] F. BITTER. The Optical Detection of Radiofrequency Resonance. *Physical Review*, **76**: 833–835, 1949. DOI: 10.1103/PhysRev.76.833 (see p. 44)

[111] R. M. WHITLEY and C. R. STROUD. Double optical resonance. *Physical Review A*, **14**: 1498–1513, 1976. DOI: 10.1103/PhysRevA.14.1498 (see p. 44)

[112] W. M. ITANO and D. J. WINELAND. „Laser Cooling and Double Resonance Spectroscopy of Stored Ions" in: *Laser Spectroscopy V, Proceedings of the Fifth International Conference*. ed. by R. W. MCKELLAR, T. OKA, and B. P. STOICHEFF Alberta, Canada: Springer-Verlag, 1981. 360 DOI: 10.1007/978-3-540-38804-3_65 (see p. 44)

[113] N. BLOEMBERGEN. Solid State Infrared Quantum Counters. *Physical Review Letters*, **2**: 84–85, 1959. DOI: 10.1103/PhysRevLett.2.84 (see p. 44)

[114] A. H. BURRELL, D. J. SZWER, S. C. WEBSTER, and D. M. LUCAS. Scalable simultaneous multiqubit readout with 99.99% single-shot fidelity. *Physical Review A*, **81**: 040302, 2010. DOI: 10.1103/PhysRevA.81.040302 (see p. 45)

[115] T. SAUTER, W. NEUHAUSER, R. BLATT, and P. TOSCHEK. Observation of quantum jumps. *Physical Review Letters*, **57**: 1696–1698, 1986. DOI: 10.1103/PhysRevLett.57.1696 (see p. 45)

[116] J. C. BERGQUIST, R. G. HULET, W. M. ITANO, and D. J. WINELAND. Observation of Quantum Jumps in a Single Atom. *Physical Review Letters*, **57**: 1699–1702, 1986. DOI: 10.1103/PhysRevLett.57.1699 (see p. 45)

[117] D. DEUTSCH and R. JOZSA. Rapid Solution of Problems by Quantum Computation. *Proceedings of the Royal Society of London. Series A: Mathematical and Physical Sciences*, **439**: 553–558, 1992. DOI: 10.1098/rspa.1992.0167 (see p. 46)

[118] S. GULDE, M. RIEBE, G. P. T. LANCASTER, C. BECHER, J. ESCHNER, H. HÄFFNER, F. SCHMIDT-KALER, I. L. CHUANG, and R. BLATT. Implementation of the Deutsch-Jozsa algorithm on an ion-trap quantum computer. *Nature*, **421**: 48–50, 2003. DOI: 10.1038/nature01336 (see p. 46)

[119] J. VON NEUMANN. „Probabilistic logics and synthesis of reliable organisms from unreliable components" in: *Automata Studies*. ed. by C. SHANNON and J. MCCARTHY Princeton University Press, 1956. 43–98 (see p. 47)

[120] C. SHANNON. A Mathematical Theory of Communication. *Bell System Technical Journal*, **27**: 379–423, 1948. DOI: 10.1002/j.1538-7305.1948.tb01338.x (see p. 47)

[121] R. W. HAMMING. Error detecting and error correcting codes. *Bell Syst. Tech. J*, **29**: 147–160, 1950. (see p. 47)

[122] A. R. CALDERBANK and P. W. SHOR. Good quantum error-correcting codes exist. *Physical Review A*, **54**: 1098–1105, 1996. DOI: 10.1103/PhysRevA.54.1098 (see p. 47)

[123] A. M. STEANE. Error Correcting Codes in Quantum Theory. *Physical Review Letters*, **77**: 793–797, 1996. DOI: 10.1103/PhysRevLett.77.793 (see p. 47)

[124] D. KIELPINSKI, V. MEYER, M. ROWE, and C. SACKETT. A decoherence-free quantum memory using trapped ions. *Science*, **291**: 1013–1015, 2001. DOI: 10.1126/science.1057357 (see p. 47)

[125] M. BROWNNUTT, M. KUMPH, P. RABL, and R. BLATT. Ion-trap measurements of electric-field noise near surfaces. *ArXiv e-prints*, 2014. arXiv: 1409.6572 [quant-ph] (see pp. 47, 50, 86, 135, 169, 170, 187)

[126] J. ALNIS, A. MATVEEV, N. KOLACHEVSKY, T. UDEM, and T. W. HÄNSCH. Subhertz linewidth diode lasers by stabilization to vibrationally and thermally compensated ultralow-expansion glass Fabry-Pérot cavities. *Physical Review A*, **77**: 053809, 2008. DOI: 10.1103/PhysRevA.77.053809 (see p. 48)

[127] S. QUINT *Formung von Laserlichtimpulsen für die Quanteninformationsverarbeitung mit $^{40}Ca^+$* 2011 (see p. 48)

[128] R. S. ABBOTT and P. J. KING. Diode-pumped Nd:YAG laser intensity noise suppression using a current shunt. *Review of Scientific Instruments*, **72**: 2001. DOI: 10.1063/1.1334627 (see p. 48)

[129] F. NOCERA. LIGO laser intensity noise suppression. *Classical and Quantum Gravity*, **21**: S481, 2004. DOI: 10.1088/0264-9381/21/5/014 (see p. 48)

[130] D. J. LARSON, J. C. BERGQUIST, J. J. BOLLINGER, W. M. ITANO, and D. J. WINELAND. Sympathetic cooling of trapped ions: A laser-cooled two-species nonneutral ion plasma. *Physical Review Letters*, **57**: 70–73, 1986. DOI: 10.1103/PhysRevLett.57.70 (see p. 49)

[131] T. A. SAVARD, K. M. O'HARA, and J. E. THOMAS. Laser-noise-induced heating in far-off resonance optical traps. *Physical Review A*, **56**: R1095–R1098, 1997. DOI: 10.1103/PhysRevA.56.R1095 (see pp. 50, 51, 187)

[132] Q. A. Turchette, B. E. King, D. Leibfried, D. M. Meekhof, C. J. Myatt, M. A. Rowe, C. A. Sackett, C. S. Wood, W. Itano, C. Monroe, and D. J. Wineland. Heating of trapped ions from the quantum ground state. *Physical Review A*, **61**: 063418, 2000. DOI: 10.1103/PhysRevA.61.063418 (see pp. 50, 83, 86, 88, 170, 193)

[133] R. Jáuregui. Nonperturbative and perturbative treatments of parametric heating in atom traps. *Physical Review A*, **64**: 053408, 2001. DOI: 10.1103/PhysRevA.64.053408 (see p. 50)

[134] R. J. Epstein, S. Seidelin, D. Leibfried, J. H. Wesenberg, J. J. Bollinger, J. M. Amini, R. B. Blakestad, J. Britton, J. Home, W. Itano, J. Jost, E. Knill, C. Langer, R. Ozeri, N. Shiga, and D. J. Wineland. Simplified motional heating rate measurements of trapped ions. *Physical Review A*, **76**: 033411, 2007. DOI: 10.1103/ PhysRevA.76.033411 (see pp. 53, 132, 170)

[135] J. H. Wesenberg, R. J. Epstein, D. Leibfried, R. B. Blakestad, J. Britton, J. Home, W. Itano, J. Jost, E. Knill, C. Langer, R. Ozeri, S. Seidelin, and D. J. Wineland. Fluorescence during Doppler cooling of a single trapped atom. *Physical Review A*, **76**: 053416, 2007. DOI: 10.1103/PhysRevA.76.053416 (see pp. 53, 132)

[136] D. Berkeland, J. Miller, J. C. Bergquist, W. Itano, and D. J. Wineland. Minimization of ion micromotion in a Paul trap. *Journal of Applied Physics*, **83**: 5025, 1998. (see pp. 55, 56, 82, 129)

[137] D. Leibfried. Individual addressing and state readout of trapped ions utilizing rf micromotion. *Physical Review A*, **60**: R3335–R3338, 1999. DOI: 10.1103/PhysRevA.60.R3335 (see pp. 55, 82)

[138] D. T. C. Allcock, T. P. Harty, H. A. Janacek, N. M. Linke, C. J. Ballance, A. M. Steane, D. M. Lucas, R. L. Jarecki, Jr, S. D. Habermehl, M. G. Blain, D. Stick, and D. L. Moehring. Micromotion compensation in a surface electrode trap by parametric excitation of trapped ions. *Applied Physics B*, **107**: 913, 2012. DOI: 10. 1007/s00340-011-4762-2 (see pp. 56, 170)

[139] D. T. C. Allcock, J. A. Sherman, D. N. Stacey, A. H. Burrell, M. J. Curtis, G. Imreh, N. M. Linke, D. J. Szwer, S. C. Webster, A. M. Steane, and D. M. Lucas. Implementation of a symmetric surface-electrode ion trap with field compensation using a modulated Raman effect. *New Journal of Physics*, **12**: 053026, 2010. DOI: 10.1088/1367-2630/12/5/053026 (see pp. 56, 170)

[140] D. P. DiVincenzo. The Physical Implementation of Quantum Computation. *Fortschritte der Physik*, **48**: 771–783, 2000. DOI: 10.1002/1521-3978(200009)48:9/11<771::AID-PROP771>3.0.CO;2-E (see pp. 56, 57)

[141] M. Harlander, R. Lechner, M. Brownnutt, R. Blatt, and W. Hänsel. Trapped-ion antennae for the transmission of quantum information. *Nature*, **471**: 200, 2011. DOI: 10.1038/nature09800 (see pp. 57, 62, 137, 170)

[142] K. R. Brown, C. Ospelkaus, Y. Colombe, A. Wilson, D. Leibfried, and D. J. Wineland. Coupled quantized mechanical oscillators. *Nature*, **471**: 196, 2011. DOI: 10. 1038/nature09721 (see pp. 57, 85, 88, 137, 170)

[143] D. KIELPINSKI, C. MONROE, and D. J. WINELAND. Architecture for a large-scale ion-trap quantum computer. *Nature*, **417**: 709, 2002. DOI: `10.1038/nature00784` (see pp. 57, 87)

[144] R. B. BLAKESTAD, C. OSPELKAUS, A. P. VANDEVENDER, J. M. AMINI, J. BRITTON, D. LEIBFRIED, and D. J. WINELAND. High-Fidelity Transport of Trapped-Ion Qubits through an X-Junction Trap Array. *Physical Review Letters*, **102**: 153002, 2009. DOI: `10.1103/PhysRevLett.102.153002` (see pp. 57, 87, 170)

[145] R. B. BLAKESTAD, C. OSPELKAUS, A. P. VANDEVENDER, J. H. WESENBERG, M. J. BIERCUK, D. LEIBFRIED, and D. J. WINELAND. Near-ground-state transport of trapped-ion qubits through a multidimensional array. *Physical Review A*, **84**: 032314, 2011. DOI: `10.1103/PhysRevA.84.032314` (see pp. 57, 170, 187)

[146] S. RITTER, C. NÖLLEKE, C. HAHN, A. REISERER, A. NEUZNER, M. UPHOFF, M. MUCKE, E. FIGUEROA, J. BOCHMANN, and G. REMPE. An elementary quantum network of single atoms in optical cavities. *Nature*, **484**: 195–200, 2012. DOI: `10.1038/nature11023` (see p. 58)

[147] B. U. BRANDSTÄTTER. *Integration of fiber mirrors and ion traps for a high-fidelity quantum interface*. PhD thesis. Universität Innsbruck, 2013. (see p. 58)

[148] P. PHAM and K. M. SVORE. A 2D Nearest-Neighbor Quantum Architecture for Factoring. *Quantum Information and Computation*, **13**: 937–962, 2013. (see p. 58)

[149] H. J. BRIEGEL, D. E. BROWNE, W. DÜR, R. RAUSSENDORF, and M. VAN DEN NEST. Measurement-based quantum computation. *Nature Physics*, **5**: 19–26, 2009. DOI: `10.1038/nphys1157` (see p. 58)

[150] H. BRIEGEL and R. RAUSSENDORF. Persistent Entanglement in Arrays of Interacting Particles. *Physical Review Letters*, **86**: 910–913, 2001. DOI: `10.1103/PhysRevLett.86.910` (see p. 58)

[151] H. F. TROTTER. On the product of semi-groups of operators. *Proceedings of the American Mathematical Society*, **10**: 545, 1959. (see p. 58)

[152] D. L. ANDREWS and D. S. BRADSHAW. Virtual photons, dipole fields and energy transfer: a quantum electrodynamical approach. *European Journal of Physics*, **25**: 845, 2004. (see p. 60)

[153] T. PRUTTIVARASIN, M. RAMM, I. TALUKDAR, A. KREUTER, and H. HÄFFNER. Trapped ions in optical lattices for probing oscillator chain models. *New Journal of Physics*, **13**: 075012, 2011. DOI: `10.1088/1367-2630/13/7/075012` (see p. 60)

[154] C. ZENER. Non-Adiabatic Crossing of Energy Levels. *Proceedings of the Royal Society A: Mathematical, Physical and Engineering Sciences*, **137**: 696–702, 1932. DOI: `10.1098/rspa.1932.0165` (see p. 64)

[155] A. S. SØRENSEN and K. MØLMER. Quantum Computation with Ions in Thermal Motion. *Physical Review Letters*, **82**: 1971–1974, 1999. DOI: `10.1103/PhysRevLett.82.1971` (see p. 66)

[156] A. C. WILSON, Y. COLOMBE, K. R. BROWN, E. KNILL, D. LEIBFRIED, and D. J. WINELAND. Tunable spin-spin interactions and entanglement of ions in separate potential wells. *Nature*, **512**: 57–60, 2014. DOI: 10.1038/nature13565 (see pp. 67, 88)

[157] M. ZHANG and L. F. WEI. Coherently manipulating cold ions in separated traps by their vibrational couplings. *Physical Review A*, **83**: 064301, 2011. DOI: 10.1103/PhysRevA.83.064301 (see p. 67)

[158] D. PORRAS and J. I. CIRAC. Effective quantum spin systems with trapped ions. *Physical Review Letters*, **92**: 207901, 2004. DOI: 10.1103/PhysRevLett.92.207901 (see p. 67)

[159] M. J. MADSEN, D. L. MOEHRING, P. MAUNZ, R. N. KOHN, JR, L. M. DUAN, and C. MONROE. Ultrafast coherent excitation of a trapped ion qubit for fast gates and photon frequency qubits. *Physical Review Letters*, **97**: 040505, 2006. DOI: 10.1103/PhysRevLett.97.040505 (see p. 67)

[160] J. H. WESENBERG. Electrostatics of surface-electrode ion traps. *Physical Review A*, **78**: 063410, 2008. DOI: 10.1103/PhysRevA.78.063410 (see pp. 68, 76, 84, 85, 112)

[161] E. Z. LIVERTS, V. B. MANDELZWEIG, and F. TABAKIN. Analytic calculation of energies and wave functions of the quartic and pure quartic oscillators. *Journal of Mathematical Physics*, **47**: 062109, 2006. DOI: 10.1063/1.2209769 arXiv: 0603165 [physics] (see p. 69)

[162] M. KUMPH *Apparatus and method for trapping charged particles and performing controlled interactions between them* 2011 (see pp. 76, 145)

[163] R. SCHMIED, J. H. WESENBERG, and D. LEIBFRIED. Optimal Surface-Electrode Trap Lattices for Quantum Simulation with Trapped Ions. *Physical Review Letters*, **102**: 233002, 2009. DOI: 10.1103/PhysRevLett.102.233002 (see pp. 76, 85, 86, 88, 145)

[164] M. CETINA, A. GRIER, J. CAMPBELL, I. L. CHUANG, and V. VULETIĆ. Bright source of cold ions for surface-electrode traps. *Physical Review A*, **76**: 041401, 2007. DOI: 10.1103/PhysRevA.76.041401 (see p. 77)

[165] T. H. KIM, P. F. HERSKIND, T. KIM, J. KIM, and I. L. CHUANG. Surface-electrode point Paul trap. *Physical Review A*, **82**: 043412, 2010. DOI: 10.1103/PhysRevA.82.043412 (see pp. 77, 88, 91)

[166] A. P. VANDEVENDER, Y. COLOMBE, J. AMINI, D. LEIBFRIED, and D. J. WINELAND. Efficient Fiber Optic Detection of Trapped Ion Fluorescence. *Physical Review Letters*, **105**: 023001, 2010. DOI: 10.1103/PhysRevLett.105.023001 (see p. 77)

[167] P. F. HERSKIND, A. DANTAN, M. ALBERT, J. P. MARLER, and M. DREWSEN. Positioning of the RF potential minimum line of a linear Paul trap with micrometer precision. *Journal of Physics B: Atomic, Molecular and Optical Physics*, **42**: 154008, 2009. DOI: 10.1088/0953-4075/42/15/154008 (see pp. 77, 89, 129)

[168] U. TANAKA, K. SUZUKI, Y. IBARAKI, and S. URABE. Design of a surface electrode trap for parallel ion strings. *Journal of Physics B: Atomic, Molecular and Optical Physics*, **47**: 035301, 2014. DOI: 10.1088/0953-4075/47/3/035301 (see p. 77)

[169] J. LABAZIEWICZ, Y. GE, P. ANTOHI, D. R. LEIBRANDT, K. R. BROWN, and I. L. CHUANG. Suppression of Heating Rates in Cryogenic Surface-Electrode Ion Traps. *Physical Review Letters*, **100**: 013001, 2008. DOI: 10.1103/PhysRevLett.100.013001 (see pp. 83, 170)

[170] M. NIEDERMAYR. *Cryogenic Surface Ion Traps*. PhD. Universität Innsbruck, 2015. (see pp. 83, 146, 175)

[171] S. SEIDELIN, J. CHIAVERINI, R. REICHLE, J. J. BOLLINGER, D. LEIBFRIED, J. BRITTON, J. H. WESENBERG, R. B. BLAKESTAD, R. J. EPSTEIN, D. B. HUME, W. ITANO, J. JOST, C. LANGER, R. OZERI, N. SHIGA, and D. J. WINELAND. Microfabricated Surface-Electrode Ion Trap for Scalable Quantum Information Processing. *Physical Review Letters*, **96**: 253003, 2006. DOI: 10.1103/PhysRevLett.96.253003 (see pp. 83, 85, 170, 178)

[172] D. R. LEIBRANDT, J. LABAZIEWICZ, R. J. CLARK, I. L. CHUANG, R. J. EPSTEIN, C. OSPELKAUS, J. H. WESENBERG, J. J. BOLLINGER, D. LEIBFRIED, D. J. WINELAND, D. STICK, J. STERK, C. MONROE, C. S. PAI, Y. LOW, R. FRAHM, and R. E. SLUSHER. Demonstration of a scalable, multiplexed ion trap for quantum information processing. *Quantum Information and Computation*, **9**: 901–919, 2009. (see pp. 84, 170)

[173] R. J. CLARK, T. LIN, K. R. BROWN, and I. L. CHUANG. A two-dimensional lattice ion trap for quantum simulation. *Journal of Applied Physics*, **105**: 013114, 2009. DOI: 10.1063/1.3056227 (see pp. 84, 88)

[174] R. ALHEIT, S. KLEINEIDAM, and F. VEDEL. Higher order non-linear resonances in a Paul trap. *International Journal of Mass Spectrometry and Ion Processes*, **154**: 155–169, 1996. DOI: 10.1016/0168-1176(96)04380-7 (see p. 84)

[175] M. HARLANDER, M. BROWNNUTT, W. HÄNSEL, and R. BLATT. Trapped-ion probing of light-induced charging effect on dielectrics. *New Journal of Physics*, **12**: 093035, 2010. DOI: 10.1088/1367-2630/12/9/093035 (see pp. 85, 88, 91, 194)

[176] J. BENHELM, G. KIRCHMAIR, C. F. ROOS, and R. BLATT. Towards fault-tolerant quantum computing with trapped ions. *Nature Physics*, **4**: 463, 2008. DOI: 10.1038/nphys961 (see p. 86)

[177] H. HÄFFNER, W. HÄNSEL, C. F. ROOS, J. BENHELM, D. CHEK-AL-KAR, M. CHWALLA, T. KÖRBER, U. D. RAPOL, M. RIEBE, P. O. SCHMIDT, C. BECHER, O. GÜHNE, W. DÜR, and R. BLATT. Scalable multiparticle entanglement of trapped ions. *Nature*, **438**: 643, 2005. DOI: 10.1038/nature04279 (see p. 86)

[178] J. M. AMINI, H. UYS, J. H. WESENBERG, S. SEIDELIN, J. BRITTON, J. J. BOLLINGER, D. LEIBFRIED, C. OSPELKAUS, A. P. VANDEVENDER, and D. J. WINELAND. Toward scalable ion traps for quantum information processing. *New Journal of Physics*, **12**: 033031, 2010. DOI: 10.1088/1367-2630/12/3/033031 (see pp. 87, 170)

[179] J. CHIAVERINI, R. B. BLAKESTAD, J. BRITTON, J. JOST, C. LANGER, D. LEIBFRIED, R. OZERI, and D. J. WINELAND. Surface-Electrode Architecture for Ion-Trap Quantum Information Processing. *Quantum Information and Computation*, **5**: 419, 2005. (see pp. 87, 88)

[180] G. HUBER, T. DEUSCHLE, W. SCHNITZLER, R. REICHLE, K. SINGER, and F. SCHMIDT-KALER. Transport of ions in a segmented linear Paul trap in printed-circuit-board technology. *New Journal of Physics*, **10**: 013004, 2008. DOI: 10.1088/1367-2630/10/1/013004 (see p. 87)

[181] C. E. PEARSON, D. R. LEIBRANDT, W. S. BAKR, W. J. MALLARD, K. R. BROWN, and I. L. CHUANG. Experimental investigation of planar ion traps. *Physical Review A*, **73**: 032307, 2006. DOI: 10.1103/PhysRevA.73.032307 (see pp. 87, 88)

[182] M. A. ROWE, A. BEN-KISH, B. DEMARCO, D. LEIBFRIED, V. MEYER, J. BEALL, J. BRITTON, J. HUGHES, W. ITANO, B. JELENKOVIĆ, C. LANGER, T. ROSENBAND, and D. J. WINELAND. Transport of quantum states and separation of ions in a dual RF ion trap. *Quantum Information and Computation*, **2**: 257, 2002. (see pp. 87, 170)

[183] A. WALTHER, F. ZIESEL, T. RUSTER, S. T. DAWKINS, K. OTT, M. HETTRICH, K. SINGER, F. SCHMIDT-KALER, and U. G. POSCHINGER. Controlling Fast Transport of Cold Trapped Ions. *Physical Review Letters*, **109**: 080501, 2012. DOI: 10.1103/PhysRevLett.109.080501 (see p. 87)

[184] R. RAUSSENDORF, D. E. BROWNE, and H. J. BRIEGEL. Measurement-based quantum computation on cluster states. *Physical Review A*, **68**: 022312, 2003. DOI: 10.1103/PhysRevA.68.022312 (see p. 88)

[185] M. HELLWIG, A. BAUTISTA-SALVADOR, K. SINGER, G. WERTH, and F. SCHMIDT-KALER. Fabrication of a planar micro Penning trap and numerical investigations of versatile ion positioning protocols. *New Journal of Physics*, **12**: 065019, 2010. DOI: 10.1088/1367-2630/12/6/065019 (see p. 88)

[186] M. J. BIERCUK, H. UYS, A. P. VANDEVENDER, N. SHIGA, W. ITANO, and J. J. BOLLINGER. High-Fidelity Quantum control using ion crystals in a penning trap. *Quantum Information and Computation*, **9**: 920, 2009. (see p. 88)

[187] J. R. CASTREJÓN-PITA, H. OHADI, D. R. CRICK, D. F. A. WINTERS, D. M. SEGAL, and R. C. THOMPSON. Novel designs for Penning traps. *Journal of Modern Optics*, **54**: 1581, 2007. DOI: 10.1080/09500340600736793 (see p. 88)

[188] C. SCHNEIDER, M. ENDERLEIN, T. HUBER, and T. SCHAETZ. Optical Trapping of an Ion. *Nature Photonics*, **4**: 772, 2010. DOI: 10.1038/nphoton.2010.236 (see p. 88)

[189] J. WELZEL, A. BAUTISTA-SALVADOR, C. ABARBANEL, V. WINEMAN-FISHER, C. WUNDERLICH, R. FOLMAN, and F. SCHMIDT-KALER. Designing spin-spin interactions with one and two dimensional ion crystals in planar micro traps. *The European Physical Journal D*, **65**: 285, 2011. DOI: 10.1140/epjd/e2011-20098-y (see p. 88)

[190] D. R. LEIBRANDT, B. YURKE, and R. E. SLUSHER. Modeling Ion Trap Thermal Noise Decoherence. *Quantum Information and Computation*, **7**: 52, 2007. (see pp. 88, 178)

[191] J. CHIAVERINI and W. E. LYBARGER. Laserless trapped-ion quantum simulations without spontaneous scattering using microtrap arrays. *Physical Review A*, **77**: 022324, 2008. DOI: 10.1103/PhysRevA.77.022324 (see pp. 88, 89)

[192] W. E. LYBARGER. *Enabling Coherent Control of Trapped Ions with Economical Multi-laser Frequency Stabilization Technology.* PhD thesis. University of California, Los Angeles, 2010. (see pp. 88, 89)

[193] T. KARIN, I. LE BRAS, A. KEHLBERGER, K. SINGER, N. DANIILIDIS, and H. HÄFFNER. Transport of charged particles by adjusting rf voltage amplitudes. *Applied Physics B: Lasers and Optics*, **106**: 117–125, 2012. DOI: 10.1007/s00340-011-4738-2 arXiv: 1011.6116 (see pp. 89, 121)

[194] T. H. KIM, P. F. HERSKIND, and I. L. CHUANG. Surface-electrode ion trap with integrated light source. *Applied Physics Letters*, **98**: 214103, 2011. DOI: 10.1063/1.3593496 (see pp. 89, 91)

[195] S. X. WANG, G. HAO LOW, N. S. LACHENMYER, Y. GE, P. F. HERSKIND, and I. L. CHUANG. Laser-induced charging of microfabricated ion traps. *Journal of Applied Physics*, **110**: 104901, 2011. DOI: 10.1063/1.3662118 (see pp. 91, 194)

[196] K. R. BROWN, R. J. CLARK, J. LABAZIEWICZ, P. RICHERME, D. R. LEIBRANDT, and I. L. CHUANG. Loading and characterization of a printed-circuit-board atomic ion trap. *Physical Review A*, **75**: 015401, 2007. DOI: 10.1103/PhysRevA.75.015401 (see pp. 91, 112)

[197] A. M. ELTONY, S. X. WANG, G. M. AKSELROD, P. F. HERSKIND, and I. L. CHUANG. Transparent ion trap with integrated photodetector. *Applied Physics Letters*, **102**: 054106, 2013. DOI: 10.1063/1.4790843 (see p. 91)

[198] M. JOHANNING, A. F. VARÓN, and C. WUNDERLICH. Quantum simulations with cold trapped ions. *Journal of Physics B: Atomic, Molecular and Optical Physics*, **42**: 154009, 2009. DOI: 10.1088/0953-4075/42/15/154009 (see p. 91)

[199] S. X. WANG, J. LABAZIEWICZ, Y. GE, R. SHEWMON, and I. L. CHUANG. Individual addressing of ions using magnetic field gradients in a surface-electrode ion trap. *Applied Physics Letters*, **94**: 094103, 2009. DOI: 10.1063/1.3095520 (see p. 91)

[200] B. P. LANYON, P. JURCEVIC, M. ZWERGER, C. HEMPEL, E. A. MARTINEZ, W. DÜR, H. J. BRIEGEL, R. BLATT, and C. F. ROOS. Measurement-Based Quantum Computation with Trapped Ions. *Phys. Rev. Lett.*, **111**: 210501, 2013. DOI: 10.1103/PhysRevLett.111.210501 (see p. 91)

[201] P. M. P. LANGEVIN. Une Formule Fondamente De Théorie Cinétique. *Annales de Chimie et de Physique*, **5**: 245, 1905. (see p. 94)

[202] D. J. WINELAND, C. MONROE, W. ITANO, D. LEIBFRIED, B. E. KING, and D. M. MEEKHOF. Experimental issues in coherent quantum-state manipulation of trapped atomic ions. *Journal of Research of the National Institute of Standards and Technology*, **103**: 259, 1998. DOI: 10.6028/jres.103.019 (see pp. 94, 177, 181, 184, 187, 194)

[203] R. W. P. DREVER, J. L. HALL, F. V. KOWALSKI, and J. HOUGH. Laser phase and frequency stabilization using an optical resonator. *Applied Physics B*, **31**: 97, 1983. DOI: 10.1007/BF00702605 (see pp. 95, 101)

[204] G. KIRCHMAIR *Frequency stabilization of a Titanium-Sapphire laser for precision spectroscopy on Calcium ions* Innsbruck, 2006 (see p. 96)

[205] M. CHWALLA. *Precision spectroscopy with* $^{40}Ca^+$ *ions in a Paul trap.* PhD thesis. Universität Innsbruck, 2009. (see pp. 96, 178)

[206] A. ERHARD *Frequenzstabilisierung eines Lasers mit 729nm Wellenlänge* 2015 (see p. 96)

[207] J. BENHELM. *Precision Spectroscopy and Quantum Information Processing with Trapped Calcium Ions.* PhD thesis. Universität Innsbruck, 2008. (see p. 101)

[208] W. W. MACALPINE and R. O. SCHILDKNECHT. Coaxial resonators with helical inner conductor. *Proceedings of the IRE*, **47**: 2099, 1959. DOI: 10.1109/JRPROC.1959.287128 (see p. 117)

[209] P. ABRIE. In: *Design of RF and microwave amplifiers and oscillators.* chap. 4, 125. Artech House Inc, 1999. (see p. 119)

[210] D. GANDOLFI *Compact RF Amplifier for Scalable Ion-Traps* 2011 (see p. 120)

[211] D. GANDOLFI, M. NIEDERMAYR, M. KUMPH, M. BROWNNUTT, and R. BLATT. Compact RF resonator for cryogenic ion traps. *Review of Scientific Instruments*, **83**: 084705, 2012. DOI: 10.1063/1.4737889 arXiv: 1204.5004 (see p. 120)

[212] C. RAAB, J. ESCHNER, J. BOLLE, H. OBERST, F. SCHMIDT-KALER, and R. BLATT. Motional sidebands and direct measurement of the cooling rate in the resonance fluorescence of a single trapped ion. *Physical Review Letters*, **85**: 538–41, 2000. (see p. 129)

[213] M. GUTIERREZ *Personal communication* 2013 (see p. 134)

[214] R. DEVOE and C. KURTSIEFER. Experimental study of anomalous heating and trap instabilities in a microscopic ^{137}Ba ion trap. *Physical Review A*, **65**: 063407, 2002. DOI: 10.1103/PhysRevA.65.063407 (see pp. 134, 170, 192, 194)

[215] N. DANIILIDIS, S. NARAYANAN, S. A. MÖLLER, R. J. CLARK, T. E. LEE, P. J. LEEK, A. WALLRAFF, S. A. SCHULZ, F. SCHMIDT-KALER, and H. HÄFFNER. Fabrication and heating rate study of microscopic surface electrode ion traps. *New Journal of Physics*, **13**: 013032, 2011. DOI: 10.1088/1367-2630/13/1/013032 (see pp. 134, 170, 178, 192, 194)

[216] E. BRAMA, A. MORTENSEN, M. KELLER, and W. LANGE. Heating rates in a thin ion trap for microcavity experiments. *Applied Physics B*, **107**: 945, 2012. DOI: 10.1007/s00340-012-5091-9 (see pp. 134, 170, 192, 194)

[217] A. HÄRTER, A. KRÜKOW, A. BRUNNER, and J. HECKER DENSCHLAG. Long-term drifts of stray electric fields in a Paul trap. *Applied Physics B*, 2013. DOI: 10.1007/s00340-013-5688-7 (see pp. 134, 192, 194)

[218] U. WARRING, C. OSPELKAUS, Y. COLOMBE, K. R. BROWN, J. M. AMINI, M. CARSJENS, D. LEIBFRIED, and D. J. WINELAND. Techniques for microwave near-field quantum control of trapped ions. *Physical Review A*, **87**: 013437, 2013. DOI: 10.1103/PhysRevA.87.013437 (see pp. 134, 170, 192, 194)

[219] J. LABAZIEWICZ, Y. GE, D. R. LEIBRANDT, S. X. WANG, R. SHEWMON, and I. L. CHUANG. Temperature dependence of electric field noise above gold surfaces. *Physical Review Letters*, **101**: 180602, 2008. DOI: 10.1103/PhysRevLett.101.180602 arXiv: 0804.2665 (see pp. 135, 170)

[220] D. J. GORMAN, P. SCHINDLER, S. SELVARAJAN, N. DANIILIDIS, and H. HÄFFNER. Two-mode coupling in a single-ion oscillator via parametric resonance. *Physical Review A*, **89**: 062332, 2014. DOI: 10.1103/PhysRevA.89.062332 (see p. 136)

[221] H. BARTZSCH, D. GLÖSS, P. FRACH, M. GITTNER, E. SCHULTHEISS, W. BRODE, and J. HARTUNG. Electrical insulation properties of sputter-deposited SiO_2 , Si_3N_4 and Al_2O_3 films at room temperature and 400 °C. *Physica Status Solidi (a)*, **206**: 514–519, 2009. DOI: 10.1002/pssa.200880481 (see p. 138)

[222] K. LANGER *Miniaturisierte zweidimensionale Ionenfallen mit segmentierter Radiofrequenzelektrode* 2013 (see pp. 138, 139)

[223] M. BUJATTI and F. N. SECHI *Microwave integrated circuit substrate including metal filled via holes and method of manufacture* 1990 (see p. 140)

[224] P. KOHLI, J. E. WHARTON, O. BRAIDE, and C. R. MARTIN. Template Synthesis of Gold Nanotubes in an Anodic Alumina Membrane. *Journal of Nanoscience and Nanotechnology*, **4**: 605–610, 2004. DOI: 10.1166/jnn.2004.015 (see p. 141)

[225] A. VLAD, M. MÁTÉFI-TEMPFLI, V. A. ANTOHE, S. FANIEL, N. RECKINGER, B. OLBRECHTS, A. CRAHAY, V. BAYOT, L. PIRAUX, S. MELINTE, and S. MÁTÉFI-TEMPFLI. Nanowire-Decorated Microscale Metallic Electrodes. *Small*, **4**: 557–560, 2008. DOI: 10.1002/smll.200700724 (see p. 141)

[226] S. MÁTÉFI-TEMPFLI, M. MÁTÉFI-TEMPFLI, A. VLAD, V. ANTOHE, and L. PIRAUX. Nanowires and nanostructures fabrication using template methods: a step forward to real devices combining electrochemical synthesis with lithographic techniques. *Journal of Materials Science: Materials in Electronics*, **20**: 249–254, 2009. DOI: 10.1007/s10854-008-9568-6 (see p. 141)

[227] D. T. C. ALLCOCK, L. GUIDONI, T. P. HARTY, C. J. BALLANCE, M. G. BLAIN, A. M. STEANE, and D. M. LUCAS. Reduction of heating rate in a microfabricated ion trap by pulsed-laser cleaning. *New Journal of Physics*, **13**: 123023, 2011. DOI: 10.1088/1367-2630/13/12/123023 (see pp. 143, 170, 178)

[228] D. A. HITE, Y. COLOMBE, A. WILSON, K. R. BROWN, U. WARRING, R. JÖRDENS, J. JOST, D. P. PAPPAS, D. LEIBFRIED, and D. J. WINELAND. 100-Fold Reduction of Electric-Field Noise in an Ion Trap Cleaned with In Situ Argon-Ion-Beam Bombardment. *Physical Review Letters*, **109**: 103001, 2012. DOI: 10.1103/PhysRevLett.109.103001 (see pp. 143, 146, 170)

[229] N. DANIILIDIS, D. J. GORMAN, L. TIAN, and H. HÄFFNER. Quantum information processing with trapped electrons and superconducting electronics. *New Journal of Physics*, **15**: 2013. DOI: 10.1088/1367-2630/15/7/073017 (see p. 143)

[230] W. LEE, E. BOSKAMP, T. GRIST, and K. KURPAD. Radiofrequency current source (RFCS) drive and decoupling technique for parallel transmit arrays using a high-power metal oxide semiconductor field-effect transistor (MOSFET). *Magnetic Resonance in Medicine*, **62**: 218–228, 2009. DOI: 10.1002/mrm.21988 (see p. 145)

[231] J. A. RICHARDS, R. M. HUEY, and J. HILLER. A new operating mode for the quadrupole mass filter. *International Journal of Mass Spectrometry and Ion Physics*, **12**: 317–339, 1973. DOI: 10.1016/0020-7381(73)80102-0 (see p. 145)

[232] M. J. SMITH *A Square-Wave-Driven Radiofrequency Quadrupole Cooler and Buncher for TITAN* 2005 (see p. 145)

[233] A. BERTON, P. TRALDI, L. DING, and F. L. BRANCIA. Mapping the Stability Diagram of a Digital Ion Trap (DIT) Mass Spectrometer Varying the Duty Cycle of the Trapping Rectangular Waveform. *Journal of the American Society for Mass Spectrometry*, **19**: 620–625, 2008. DOI: 10.1016/j.jasms.2007.12.012 (see p. 145)

[234] M. H. OLIVEIRA and J. A. MIRANDA. Biot-Savart-like law in electrostatics. *European Journal of Physics*, **22**: 31–38, 2001. DOI: 10.1088/0143-0807/22/1/304 (see p. 145)

[235] R. SCHMIED. Electrostatics of gapped and finite surface electrodes. *New Journal of Physics*, **12**: 023038, 2010. DOI: 10.1088/1367-2630/12/2/023038 (see p. 145)

[236] R. SCHMIED, J. H. WESENBERG, and D. LEIBFRIED. Quantum simulation of the hexagonal Kitaev model with trapped ions. *New Journal of Physics*, **13**: 115011, 2011. DOI: 10.1088/1367-2630/13/11/115011 arXiv: 1107.0181 (see p. 145)

[237] T. FÖRSTER. Zwischenmolekulare Energiewanderung und Fluoreszenz. *Annalen der Physik*, **437**: 55–75, 1948. DOI: 10.1002/andp.19484370105 (see p. 145)

[238] R. HILDNER, D. BRINKS, J. B. NIEDER, R. J. COGDELL, and N. F. VAN HULST. Quantum Coherent Energy Transfer over Varying Pathways in Single Light-Harvesting Complexes. *Science*, **340**: 1448–1451, 2013. DOI: 10.1126/science.1235820 (see p. 145)

[239] M. NIEDERMAYR, K. LAKHMANSKIY, M. KUMPH, S. PARTEL, J. EDLINGER, M. BROWNNUTT, and R. BLATT. Cryogenic surface ion trap based on intrinsic silicon. *New Journal of Physics*, **16**: 113068, 2014. DOI: 10.1088/1367-2630/16/11/113068 (see pp. 146, 170)

[240] N. DANIILIDIS, S. GERBER, G. BOLLOTEN, M. RAMM, A. RANSFORD, E. ULIN-AVILA, I. TALUKDAR, and H. HÄFFNER. Surface noise analysis using a single-ion sensor. *Phys. Rev. B*, **89**: 245435, 2014. DOI: 10.1103/PhysRevB.89.245435 (see pp. 146, 170)

[241] L. SLODIČKA, G. HÉTET, N. RÖCK, P. SCHINDLER, M. HENNRICH, and R. BLATT. Atom-Atom Entanglement by Single-Photon Detection. *Physical Review Letters*, **110**: 083603, 2013. DOI: 10.1103/PhysRevLett.110.083603 (see p. 147)

[242] L. DESLAURIERS, S. OLMSCHENK, D. STICK, W. K. HENSINGER, J. STERK, and C. MONROE. Scaling and Suppression of Anomalous Heating in Ion Traps. *Physical Review Letters*, **97**: 103007, 2006. DOI: 10.1103/PhysRevLett.97.103007 (see pp. 169, 170, 181–183, 194)

[243] C. OSPELKAUS, U. WARRING, Y. COLOMBE, K. R. BROWN, J. M. AMINI, D. LEIBFRIED, and D. J. WINELAND. Microwave quantum logic gates for trapped ions. *Nature*, **476**: 181–4, 2011. DOI: 10.1038/nature10290 (see p. 170)

[244] G. VITTORINI, K. WRIGHT, K. R. BROWN, A. W. HARTER, and S. C. DORET. Modular cryostat for ion trapping with surface-electrode ion traps. *The Review of Scientific Instruments*, **84**: 043112, 2013. DOI: 10.1063/1.4802948 (see p. 170)

[245] B. E. KING, C. S. WOOD, C. J. MYATT, Q. A. TURCHETTE, D. LEIBFRIED, W. ITANO, C. MONROE, and D. J. WINELAND. Cooling the Collective Motion of Trapped Ions to Initialize a Quantum Register. *Physical Review Letters*, **81**: 1525, 1998. DOI: 10.1103/PhysRevLett.81.1525 (see p. 170)

[246] S. A. SCHULZ, U. G. POSCHINGER, F. ZIESEL, and F. SCHMIDT-KALER. Sideband cooling and coherent dynamics in a microchip multi-segmented ion trap. *New Journal of Physics*, **10**: 045007, 2008. DOI: 10.1088/1367-2630/10/4/045007 (see p. 170)

[247] P. F. HERSKIND, S. X. WANG, M. SHI, Y. GE, M. CETINA, and I. L. CHUANG. Microfabricated surface ion trap on a high-finesse optical mirror. *Optics Letters*, **36**: 3045–7, 2011. (see p. 170)

[248] C. L. ARRINGTON, K. S. MCKAY, E. D. BACA, J. J. COLEMAN, Y. COLOMBE, P. FINNEGAN, D. A. HITE, A. E. HOLLOWELL, R. JÖRDENS, J. D. JOST, D. LEIBFRIED, A. M. ROWEN, U. WARRING, M. WEIDES, A. C. WILSON, D. J. WINELAND, and D. P. PAPPAS. Micro-fabricated stylus ion trap. *Review of Scientific Instruments*, **84**: 085001, 2013. DOI: 10.1063/1.4817304 (see p. 170)

[249] J. BENHELM, G. KIRCHMAIR, C. F. ROOS, and R. BLATT. Experimental quantum-information processing with ^{43}Ca^{+} ions. *Physical Review A*, **77**: 062306, 2008. DOI: 10.1103/PhysRevA.77.062306 (see p. 170)

[250] E. MOUNT, S.-Y. BAEK, M. BLAIN, D. STICK, D. GAULTNEY, S. CRAIN, R. NOEK, T. KIM, P. MAUNZ, and J. KIM. Single qubit manipulation in a microfabricated surface electrode ion trap. *New Journal of Physics*, **15**: 093018, 2013. DOI: 10.1088/1367-2630/15/9/093018 (see p. 170)

[251] C. J. MYATT, B. E. KING, Q. A. TURCHETTE, C. A. SACKETT, D. KIELPINSKI, W. ITANO, C. MONROE, and D. J. WINELAND. Decoherence of quantum superpositions through coupling to engineered reservoirs. *Nature*, **403**: 269–73, 2000. DOI: 10.1038/35002001 (see p. 170)

[252] D. A. HITE, Y. COLOMBE, A. WILSON, D. T. C. ALLCOCK, D. LEIBFRIED, D. J. WINELAND, and D. P. PAPPAS. Surface science for improved ion traps. *MRS Bulletin*, **38**: 826–833, 2013. DOI: 10.1557/mrs.2013.207 (see p. 170)

[253] C. TAMM, D. ENGELKE, and V. BÜHNER. Spectroscopy of the electric-quadrupole transition $^2S_{1/2}(F = 0)-^2D_{3/2}(F = 2)$ in trapped ^{171}Yb^{+}. *Physical Review A*, **61**: 053405, 2000. DOI: 10.1103/PhysRevA.61.053405 (see p. 170)

[254] J. BRITTON. *Microfabrication techniques for trapped ion quantum information processing*. PhD thesis. University of Colorado, 2008. (see p. 170)

[255] J. CHIAVERINI and J. M. SAGE. Insensitivity of the rate of ion motional heating to trap-electrode material over a large temperature range. *Physical Review A*, **89**: 012318, 2014. DOI: 10.1103/PhysRevA.89.012318 (see p. 170)

[256] U. G. POSCHINGER, G. HUBER, F. ZIESEL, M. DEISS, M. HETTRICH, S. A. SCHULZ, K. SINGER, G. POULSEN, M. DREWSEN, R. J. HENDRICKS, and F. SCHMIDT-KALER. Coherent manipulation of a 40 Ca + spin qubit in a micro ion trap. *Journal of Physics B: Atomic, Molecular and Optical Physics*, **42**: 154013, 2009. DOI: 10.1088/0953-4075/42/15/154013 (see p. 170)

[257] N. AKERMAN, Y. GLICKMAN, S. KOTLER, A. KESELMAN, and R. OZERI. Quantum control of ^{88}Sr^{+} in a miniature linear Paul trap. *Applied Physics B*, **107**: 1167, 2012. DOI: 10.1007/s00340-011-4807-6 (see p. 170)

[258] F. SCHMIDT-KALER, C. ROOS, H. C. NÄGERL, H. ROHDE, S. GULDE, A. MUNDT, M. LEDERBAUER, G. THALHAMMER, T. ZEIGER, P. BARTON, L. HORNEKAER, G. REYMOND, D. LEIBFRIED, J. ESCHNER, and R. BLATT. Ground state cooling, quantum state engineering and study of decoherence of ions in Paul traps. *Journal of Modern Optics*, **47**: 2573–2582, 2000. DOI: 10.1080/095003400750039555 (see p. 170)

[259] M. HARLANDER. *Architecture for a scalable ion-trap quantum computer*. PhD thesis. Universität Innsbruck, 2012. (see p. 170)

[260] H. ROHDE, S. T. GULDE, C. ROOS, P. A. BARTON, D. LEIBFRIED, J. ESCHNER, F. SCHMIDT-KALER, and R. BLATT. Sympathetic ground-state cooling and coherent manipulation with two-ion crystals. *Journal of Physics B: Atomic, Molecular and Optical Physics*, **3**: S34, 2001. DOI: 10.1088/1464-4266/3/1/357 (see pp. 170, 178)

[261] J. BRITTON, D. LEIBFRIED, J. A. BEALL, R. B. BLAKESTAD, J. H. WESENBERG, and D. J. WINELAND. Scalable arrays of RF Paul traps in degenerate Si. *Applied Physics Letters*, **95**: 173102, 2009. DOI: 10.1063/1.3254188 (see p. 170)

[262] S. X. WANG. *Quantum Gates, Sensors, and Systems with Trapped Ions*. PhD thesis. Massachusetts Institute of Technology, 2012. (see p. 170)

[263] M. KUMPH, P. HOLZ, K. LANGER, M. NIEDERMAYR, M. BROWNNUTT, and R. BLATT. Operation of a planar-electrode ion trap array with adjustable RF electrodes. *ArXiv e-prints*, 2014. arXiv: 1402.0791 [quant-ph] (see p. 170)

[264] S. C. DORET, J. M. AMINI, K. WRIGHT, C. VOLIN, T. KILLIANX, A. OZAKIN, D. DENISON, H. HAYDEN, C. S. PAI, R. E. SLUSHER, and A. W. HARTER. Controlling trapping potentials and stray electric fields in a microfabricated ion trap through design and compensation. *New Journal of Physics*, **14**: 073012, 2012. DOI: 10.1088/1367-2630/14/7/073012 (see p. 170)

[265] K. MCKAY, D. HITE, Y. COLOMBE, R. JÖRDENS, A. C. WILSON, D. H. SLICHTER, D. T. C. ALLCOCK, D. LEIBFRIED, D. J. WINELAND, and D. P. PAPPAS. Ion-trap electrode preparation with Ne [+] bombardment. *arXiv preprint*, 1–4, 2014. arXiv: arXiv: 1406.1778v1 (see p. 170)

[266] G. WILPERS, P. SEE, P. GILL, and A. G. SINCLAIR. A monolithic array of three-dimensional ion traps fabricated with conventional semiconductor technology. *Nature Nanotechnology*, **7**: 572, 2012. DOI: 10.1038/nnano.2012.126 (see p. 170)

[267] K. K. MEHTA, A. M. ELTONY, C. D. BRUZEWICZ, I. L. CHUANG, R. J. RAM, J. M. SAGE, and J. CHIAVERINI. Ion traps fabricated in a CMOS foundry. *Applied Physics Letters*, **105**: 21–24, 2014. DOI: 10.1063/1.4892061 arXiv: 1406.3643 (see p. 170)

[268] L. DESLAURIERS, P. HALJAN, P. J. LEE, K.-A. BRICKMAN, B. BLINOV, M. J. MADSEN, and C. MONROE. Zero-point cooling and low heating of trapped ^{111}Cd$^+$ ions. *Physical Review A*, **70**: 043408, 2004. DOI: 10.1103/PhysRevA.70.043408 (see p. 170)

[269] J. F. GOODWIN, G. STUTTER, R. C. THOMPSON, and D. M. SEGAL. Sideband cooling an ion to the quantum ground state in a Penning trap with very low heating rate. *ArXiv*, 4, 2014. arXiv: 1407.6121 (see p. 170)

[270] J. HOME. *Entanglement of Two Trapped-Ion Spin Qubits.* PhD thesis. Oxford University, 2006. (see p. 170)

[271] S. X. WANG, J. LABAZIEWICZ, Y. GE, R. SHEWMON, and I. L. CHUANG. Demonstration of a quantum logic gate in a cryogenic surface-electrode ion trap. *Physical Review A*, **81**: 062332, 2010. (see p. 170)

[272] G. POULSEN, Y. MIROSHNYCHENKO, and M. DREWSEN. Efficient ground-state cooling of an ion in a large room-temperature linear Paul trap with a sub-Hertz heating rate. *Physical Review A*, **86**: 051402, 2012. DOI: 10.1103/PhysRevA.86.051402 (see p. 170)

[273] N. D. GUISE, S. D. FALLEK, H. HAYDEN, C.-S. PAI, C. VOLIN, K. R. BROWN, J. T. MERRILL, A. W. HARTER, J. M. AMINI, L. M. LUST, K. MULDOON, D. CARLSON, and J. BUDACH. In-Vacuum Active Electronics for Microfabricated Ion Traps. *Review of Scientific Instruments*, **85**: 063101, 2014. DOI: 10.1063/1.4879136 arXiv: 1403.3662 (see p. 170)

[274] D. STICK, W. K. HENSINGER, S. OLMSCHENK, M. J. MADSEN, K. SCHWAB, and C. MONROE. Ion trap in a semiconductor chip. *Nature Physics*, **2**: 36–39, 2006. DOI: 10.1038/nphys171 (see p. 170)

[275] S. X. WANG, Y. GE, J. LABAZIEWICZ, E. DAULER, K. BERGGREN, and I. L. CHUANG. Superconducting microfabricated ion traps. *Applied Physics Letters*, **97**: 244102, 2010. DOI: 10.1063/1.3526733 (see p. 170)

[276] C. D. BRUZEWICZ, J. M. SAGE, and J. CHIAVERINI. Measurement of ion motional heating rates over a range of trap frequencies and temperatures. *Physical Review A*, **91**: 041402, 2015. DOI: 10.1103/PhysRevA.91.041402 (see p. 170)

[277] J. J. MCLOUGHLIN, A. H. NIZAMANI, J. D. SIVERNS, R. C. STERLING, M. D. HUGHES, B. LEKITSCH, B. STEIN, S. WEIDT, and W. K. HENSINGER. Versatile ytterbium ion trap experiment for operation of scalable ion-trap chips with motional heating and transition-frequency measurements. *Physical Review A*, **83**: 013406, 2011. DOI: 10.1103/PhysRevA.83.013406 (see p. 170)

[278] M. STEINER, H. M. MEYER, C. DEUTSCH, J. REICHEL, and M. KÖHL. Single Ion Coupled to an Optical Fiber Cavity. *Physical Review Letters*, **110**: 043003, 2013. DOI: 10.1103/PhysRevLett.110.043003 (see p. 170)

[279] S. WEIDT, J. RANDALL, S. WEBSTER, E. STANDING, A. RODRIGUEZ, A. WEBB, B. LEKITSCH, and W. HENSINGER. Ground-state cooling of a trapped ion using long-wavelength radiation. *ArXiv e-prints*, 2015. arXiv: 1501.01562 [quant-ph] (see p. 170)

[280] D. T. C. ALLCOCK, T. P. HARTY, C. J. BALLANCE, B. C. KEITCH, N. M. LINKE, D. N. STACEY, and D. M. LUCAS. A microfabricated ion trap with integrated microwave circuitry. *Applied Physics Letters*, **102**: 044103, 2013. DOI: 10.1063/1.4774299 (see p. 170)

[281] R. MCCONNELL, C. BRUZEWICZ, J. CHIAVERINI, and J. SAGE. Reduction of trapped ion anomalous heating by in situ surface plasma cleaning. *ArXiv e-prints*, 2015. arXiv: 1505.03844 [physics.atom-ph] (see p. 170)

[282] V. LETCHUMANAN, G. WILPERS, M. BROWNNUTT, P. GILL, and A. G. SINCLAIR. Zero-point cooling and heating-rate measurements of a single ^{88}Sr$^+$ ion. *Physical Review A*, **75**: 063425, 2007. DOI: 10.1103/PhysRevA.75.063425 (see p. 170)

[283] L. D. LANDAU and E. M. LIFSHITZ. *Statistical Physics 3rd Edition Part 1*. Elvesier, 1980. (see p. 173)

[284] D. GRIFFITHS. *Introduction to Electrodynamics*. Prentice Hall, 1999. (see p. 173)

[285] W. M. SIEBERT. *Circuits, Signals, and Systems*. The MIT Press, 1986. (see p. 174)

[286] C. HENKEL, S. PÖTTING, and M. WILKENS. Loss and heating of particles in small and noisy traps. *Applied Physics B*, **69**: 379–387, 1999. DOI: 10.1007/s003400050823 (see pp. 174, 181)

[287] M. PARROT, U. S. INAN, N. G. LEHTINEN, and J. L. PINÇON. Penetration of lightning MF signals to the upper ionosphere over VLF ground-based transmitters. *Journal of Geophysical Research*, **114**: A12318, 2009. DOI: 10.1029/2009JA014598 (see p. 174)

[288] N. L. BUNCH and J. LABELLE. Fully resolved observations of auroral medium frequency burst radio emissions. *Geophysical Research Letters*, **36**: L15104, 2009. DOI: 10.1029/2009GL038513 (see p. 174)

[289] D. HANSEN. „Review of EMC main aspects in fast PLC including some history" in: *Proceedings of the IEEE International Symposium on Electromagnetic Compatibility*. 2003. 184 DOI: 10.1109/ICSMC2.2003.1428226 (see p. 174)

[290] ITU. Recommendation ITU-R P.372-10. *International Telecommunications Union*, 2009. (see p. 174)

[291] I. FERNÁNDEZ, P. ANGUEIRA, I. LANDA, A. ARRINDA, D. GUERRA, and U. GIL. Indoor noise measurements in medium wave band. *Electronics Letters*, **46**: 1162, 2010. DOI: 10.1049/el.2010.1629 (see p. 174)

[292] I. LANDA, A. ARRINDA, I. FERNÁNDEZ, and P. ANGUEIRA. Indoor Radio Noise Long-Term Measurements in Medium Wave Band in Buildings of City Areas in the North of Spain. *IEEE Antennas and Wireless Propation Letters*, **10**: 17–20, 2011. DOI: 10.1109/LAWP.2011.2107872 (see p. 174)

[293] K. R. BROWN, A. WILSON, Y. COLOMBE, C. OSPELKAUS, A. M. MEIER, E. KNILL, D. LEIBFRIED, and D. J. WINELAND. Single-qubit-gate error below 10^{-4} in a trapped ion. *Physical Review A*, **84**: 030303, 2011. DOI: 10.1103/PhysRevA.84.030303 (see p. 175)

[294] M. KANDA. Standard Probes for Electromagnetic Field Measurements. *IEEE Transactions on Antennas and Propagation*, **41**: 1349, 1993. DOI: 10.1109/8.247775 (see p. 176)

[295] J. B. JOHNSON. Thermal Agitation of Electricity in Conductors. *Physical Review*, **32**: 110, 1928. DOI: 10.1103/PhysRev.32.110 (see p. 177)

[296] H. NYQUIST. Thermal Agitation of Electric Charge in Conductors. *Physical Review*, **32**: 110, 1928. DOI: 10.1103/PhysRev.32.110 (see p. 177)

[297] S. K. LAMOREAUX. Thermalization of trapped ions: A quantum perturbation approach. *Physical Review A*, **56**: 4970–4975, 1997. DOI: 10.1103/PhysRevA.56.4970 (see pp. 177, 184)

[298] J. L. ZAR. Measurement of Low Resistance and the AC Resistance of Superconductors. *Review of Scientific Instruments*, **34**: 801, 1963. DOI: 10.1063/1.1718579 (see p. 178)

[299] L. K. J. VANDAMME, A. KHALFALLAOUI, G. LEROY, and G. VÉLU. Thermal equilibrium noise with 1/f spectrum from frequency independent dielectric losses in barium strontium titanate. *Journal of Applied Physics*, **107**: 053717, 2010. DOI: 10.1063/1.3327446 (see p. 178)

[300] A. G. SINCLAIR, M. A. WILSON, and P. GILL. Improved three-dimensional control of a single strontium ion in an endcap trap. *Optics Communications*, **190**: 193–203, 2001. DOI: 10.1016/S0030-4018(01)01057-4 (see p. 178)

[301] Y. SHLEPNEV and S. McMORROW. Nickel characterization for interconnect analysis. *IEEE International Symposium on EMC*, 524, 2011. DOI: 10.1109/ISEMC.2011.6038368 (see p. 179)

[302] H. PORITSKY and R. P. JERRARD. Eddy-current losses in a semi-infinite solid due to a nearby alternating current. *Transactions of the American Institute of Electrical Engineers, Part I: Communication and Electronics*, **73**: 97–106, 1954. DOI: 10.1109/TCE.1954.6372119 (see p. 179)

[303] J. LEKNER. Electrostatics of two charged conducting spheres. *Proceedings of the Royal Society A: Mathematical, Physical and Engineering Sciences*, **468**: 2829–2848, 2012. DOI: 10.1098/rspa.2012.0133 (see p. 181)

[304] G. CHEN and A. E. DAVIES. Electric stress computation - a needle-plane electrode system with space charge effects. *COMPEL*, **15**: 40, 1996. DOI: 10.1108/03321649610120679 (see p. 182)

[305] J. W. EKIN. *Experimental Techniques for Low-Temperature measurements*. New York: Oxford University Press, 2007. (see p. 184)

[306] B. E. BLAIR. „Heterodyne Techniques" in: *Time and Frequency: Theory and Fundamentals*. Boulder: US Dept of Commerce National Bureau of Standards, 1974. 162 (see p. 185)

[307] C. GREBENKEMPER. Local oscillator phase noise and its effect on receiver performance. *Watkins-Johnson Company Tech-notes*, **8**: 1981. (see p. 185)

[308] C. E. CALOSSO, Y. GRUSON, and E. RUBIOLA. „Phase noise and amplitude noise in DDS" in: *Frequency Control Symposium (FCS), 2012 IEEE International*. 2012. 1–6 DOI: 10.1109/FCS.2012.6243619 (see p. 186)

[309] H. G. DEHMELT. Radiofrequency Spectroscopy of Stored Ions II: Spectroscopy. *Advances in Atomic and Molecular Physics*, **5**: 109–154, 1969. DOI: 10.1016/S0065-2199(08)60156-6 (see p. 187)

[310] S. MARVIN. 1/f noise. *Proceedings of the IEEE*, **70**: 212–218, 1982. DOI: 10.1109/PROC.1982.12282 (see p. 188)

[311] J. B. JOHNSON. The Schottky Effect in Low Frequency Circuits. *Physical Review*, **26**: 71–85, 1925. DOI: 10.1103/PhysRev.26.71 (see p. 188)

[312] F. N. HOOGE. 1/f noise is no surface effect. *Physics Letters A*, **29**: 139–140, 1969. DOI: 10.1016/0375-9601(69)90076-0 (see p. 188)

[313] P. DUTTA and P. M. HORN. Low-frequency fluctuations in solids: 1/f noise. *Reviews of Modern Physics*, **53**: 497, 1981. DOI: 10.1103/RevModPhys.53.497 (see p. 188)

[314] A. K. RAYCHAUDHURI. Measurement of 1 / f noise and its application in materials science. *Current Opinion in Solid State and Materials Science*, **6**: 67–85, 2002. DOI: 10.1016/S1359-0286(02)00025-6 (see p. 188)

[315] H. SUTCLIFFE. Current-induced resistor noise not attributable entirely to fluctuations of conductivity. *Electronics Letters*, **7**: 160–161, 1971. DOI: 10.1049/el:19710104 (see p. 189)

[316] J. H. J. LORTEIJE and A. M. H. HOPPENBROUWERS. Amplitude modulation by $1/f$ noise in resistors results in $1/\Delta f$ noise. *Philips Research Reports*, **26**: 29, 1971. (see pp. 189, 192)

[317] D. A. BELL. Harmonic generation in 1/f noise (resistors). *Journal of Physics D: Applied Physics*, **9**: L127, 1976. DOI: 10.1088/0022-3727/9/11/002 (see p. 192)

[318] B. K. JONES. $1/f$ and $1/\Delta f$ noise produced by a radio-frequency current in a carbon resistor. *Electronics Letters*, **12**: 110–111, 1976. DOI: 10.1049/el:19760086 (see p. 192)

[319] G. J. M. VAN HELVOORT and H. G. E. BECK. Model for the excitation of 1/f noise by high-frequency A.C. signals. *Electronics Letters*, **13**: 542–544, 1977. DOI: 10.1049/el:19770390 (see p. 192)

[320] R. H. M. CLEVERS. 1/f noise in ion-implanted resistors between 77 and 300 K. *Journal of Applied Physics*, **62**: 1877, 1987. DOI: 10.1063/1.339572 (see p. 192)

[321] C. HERRING and M. NICHOLS. Thermionic Emission. *Reviews of Modern Physics*, **21**: 185, 1949. DOI: 10.1103/RevModPhys.21.185 (see p. 193)

[322] C. SANDOGHDAR, C. SUKENIK, E. A. HINDS, and S. HAROCHE. Direct measurement of the van der Waals interaction between an atom and its images in a micron-sized cavity. *Physical Review Letters*, **68**: 3432, 1992. DOI: 10.1103/PhysRevLett.68.3432 (see p. 193)

[323] D. HARBER, J. OBRECHT, J. MCGUIRK, and E. CORNELL. Measurement of the Casimir-Polder force through center-of-mass oscillations of a Bose-Einstein condensate. *Physical Review A*, **72**: 033610, 2005. DOI: 10.1103/PhysRevA.72.033610 (see p. 193)

[324] J. B. CAMP, T. W. DARLING, and R. E. BROWN. Macroscopic variations of surface potentials of conductors. *Journal of Applied Physics*, **69**: 7126–7129, 1991. DOI: 10.1063/1.347601 (see p. 193)

[325] T. W. DARLING, G. I. OPAT, F. ROSSI, and G. F. MOORHEAD. The fall of charged particles under gravity: A study of experimental problems. *Reviews of Modern Physics*, **64**: 237, 1992. DOI: 10.1103/RevModPhys.64.237 (see p. 193)

[326] R. DUBESSY, T. COUDREAU, and L. GUIDONI. Electric field noise above surfaces: A model for heating-rate scaling law in ion traps. *Physical Review A*, **80**: 031402, 2009. DOI: 10.1103/PhysRevA.80.031402 (see p. 193)

[327] G. H. LOW, P. F. HERSKIND, and I. L. CHUANG. Finite-geometry models of electric field noise from patch potentials in ion traps. *Physical Review A*, **84**: 053425, 2011. DOI: 10.1103/PhysRevA.84.053425 (see p. 193)

[328]   J. B. CAMP, T. W. DARLING, and R. E. BROWN. Effect of crystallites on surface potential variations of Au and graphite. *Journal of Applied Physics*, **71**: 783–785, 1992. DOI: 10.1063/1.351358 (see p. 194)

[329]   F. ROSSI and G. I. OPAT. Observations of the effects of adsorbates on patch potentials. *Journal of Physics D: Applied Physics*, **25**: 1349–1353, 2000. DOI: 10.1088/0022-3727/25/9/012 (see p. 194)

[330]   J. OBRECHT, R. WILD, and E. CORNELL. Measuring electric fields from surface contaminants with neutral atoms. *Physical Review A*, **75**: 062903, 2007. DOI: 10.1103/PhysRevA.75.062903 (see p. 194)

[331]   E. L. MURPHY and R. H. GOOD. Thermionic Emission, Field Emission, and the Transition Region. *Physical Review*, **102**: 1464–1473, 1956. DOI: 10.1103/PhysRev.102.1464 (see p. 194)

[332]   L. B. LINFORD. Recent Developments in the Study of the External Photoelectric Effect. *Reviews of Modern Physics*, **5**: 34, 1933. DOI: 10.1103/RevModPhys.5.34 (see p. 194)

[333]   W. SCHOTTKY. Über spontane Stromschwankungen in verschiedenen Elektrizitätsleitern. *Annalen der Physik*, **362**: 541–567, 1918. DOI: 10.1002/andp.19183622304 (see p. 195)

[334]   M. TRINGIDES and R. GOMER. Diffusion anisotropy of oxygen and of tungsten on the tungsten (211) plane. *The Journal of Chemical Physics*, **84**: 4049, 1986. DOI: 10.1063/1.450066 (see p. 195)

[335]   C. V. DHARMADHIKARI, R. S. KHAIRNAR, and D. S. JOAG. Noise in field-induced electron emission from graphite composite: spectral density and autocorrelation investigations. *Journal of Physics D: Applied Physics*, **24**: 1842, 1991. DOI: 10.1088/0022-3727/24/10/019 (see p. 195)

[336]   F. LEFERINK, F. SILVA, J. CATRYSSE, S. BATTERMAN, and V. BEAUVOIS. Radio Science Bulletin. *International Union of Radio Science*, **334**: 2010. (see p. 196)

[337]   W. M. GREENE, J. B. KRUGER, and G. KOOI. Magnetron etching of polysilicon: Electrical damage. *Journal of Vacuum Science & Technology B: Microelectronics and Nanometer Structures*, **9**: 366, 1991. DOI: 10.1116/1.585577 (see p. 200)

[338]   S. FANG, J. P. MCVITTIE, and S. MEMBER. Thin-Oxide Damage from Gate Charging During Plasma Processing. *Electron Device Letters, IEEE*, **13**: 288–290, 1992. DOI: 10.1109/55.145056 (see p. 200)